FROM NUMBERS TO FRACTALS

A short introduction
to mathematics

Kurt Martin

Copyright © 2014 Kurt Martin, Red Horse, Munich
Red Horse are: Peter Hauser, Kurt Martin and Jack Eden
ISBN: 978-1505296273
Email: Red.Horse@gmx.net
All rights reserved.

1. INTRODUCTION 5

2. NUMBERS 9

2.1. Natural Numbers 9
2.2. Rational Numbers 13
2.3. Real Numbers 18
2.4. Complex Numbers 23
2.5. The Rule of Three 30

3. FUNCTIONS 36

3.1. Mappings 36
3.2. Exponential Function and Logarithm 39
3.3. Trigonometric Functions 48
3.4. Euler's Formula 56

4. ANALYSIS 63

4.1. The Limit 63
4.2. Differential Calculus 68
4.3. Integral Calculus 82
4.4. Differential Equations 90

5. STOCHASTIC   98
5.1. Statistics   98
5.2. Calculus of Probabilities   110

6. ANALYTIC GEOMETRY   129
6.1. Euclidean and non-Euclidean Geometry   129
6.2. Vectors   136
6.3. Vector Analysis   152
6.4. Matrices   159

7. FRACTAL GEOMETRY   182

1. Introduction

Mathematics is one of the cornerstones of human civilization. But even though everyone has to learn mathematics at school and is using it practically every day, it still seems to be so strange, complicated, and abstract as if it hadn't anything to do with our life. The sciences of biology or history seem to be so much closer to our daily life, after all they tell us about things that happen in the lives of humans or animals.
And it is true that mathematics has no connection to our world, even though many fields of mathematics were inspired by problems that natural scientists dealt with. Social sciences try to explain the world humans live in, and natural sciences try to detect the basic rules and laws that every object in the universe has to obey. But mathematics in its pure form has no relation to our world. Mathematics doesn't care about the world around us; it lives in a world of its own.
This can easily be understood if we take a look at the objects that mathematics deals with. These objects are not objects of our world, but idealized pictures that do not exist in our world. Let's take the example of a triangle. This is a rather simple object, and triangles as known from mathematics, objects with three sides and three angles, unquestionable seem to exist in our world. We have no problem to draw a triangle. But the triangle that we draw is not the triangle mathematics deals with.
A triangle is formed by three sides. That seems to be easy enough, but a side, a line in mathematics is a "breadthless length", as the Greek mathematician Euclid wrote in the third century B.C. in his ground-breaking work about

geometry. A line has no width. And even when we use the sharpest pencil, we are not able to draw a line with no width. The line might be narrow, but it would still have a width.

Similarly, the numbers that mathematicians use for calculation simply don't exist in our world. One, three or five as numbers don't mean anything. We might find a length of five meters, a volume of three liters, or a show with one viewer (which would be a deplorable incident for the actors). In our world we can only find physical properties, i.e. numbers with a unit. A number without a unit is without any physical meaning.

You could assume that mathematics is a science that abstracts the physical reality around us, takes away the width of lines and units of numbers to provide pure elements that you can easily calculated with. And, in fact, that's how mathematics was derived in the first place: People started to calculate because they wanted to know the number of slaves they owned, the volume of the harvest gathered, or the size of a property. In later times, humans wanted to predict the course of the heavenly bodies to assure that religious festivities would be done at the right time. Geometry and algebra were developed because humans had concrete problems they wanted to solve.

But mathematics didn't stop there. Mathematicians didn't only want to show that a triangle whose sides have lengths of 3, 4, and 5 units possess a right angle opposite to the side with a length of five units; mathematicians wanted to show that for *all* triangles which have a right angle the sum of the square of the two sides adjacent to the right angle is equal to the square of the side opposite to the right angle (which is known as the Pythagorean Theorem).

Over time, mathematics developed into a science that started with some axioms, some assumption that seemed to be obviously true, and tried to derive and prove any possible connections between the objects of its world, the numbers and geometrical objects. And mathematicians didn't care if their objects had any equivalence in the real world, as long as their proofs worked within the realm of their objects and rules. Mathematicians might be happy if physicists pass by, take a look at the collection of mathematical formulas and pick one to describe a problem in the real world. But describing the real world is not intended by mathematicians.

There's another example to show this. It is again taken from geometry. The geometry that we learn at school is called Euclidean geometry. It is the geometry of a flat surface where the sum of all angles in a triangle is 180°. But you could also imagine a geometry where the sum of all angles in a triangle is less than 180° or another geometry where this sum is even bigger (we will have a look at them later on).

Mathematicians were playing around with these strange geometries already in the 19th century. But nobody could imagine a practical use for them. This changed when Albert Einstein presented his General Theory of Relativity at the beginning of the 20th century. This theory describes gravity as deformations of the space. Space is not flat as Euclid assumed, but is formed in various ways. All of a sudden, there was a practical use for these funny, nun-Euclidean geometries. Mathematics never intended this.

This short introduction into mathematics will give an overview about mathematical topics that found their use in natural science. And it describes some discoveries that we consider self-evident today, but that were unknown for centuries, like the discovery of the zero or the fact the some numbers like π cannot be written as a fraction. Other discoveries of mathematics still seem to be strange like the "non-integer" dimensions that are used to describe fractals. But natural science even found some use for this odd fractal geometry.

This short introduction to mathematics is not meant to be a comprehensive textbook. It only targets to explain important terms like "vector", "matrix", and "integral" and how to use them.

2. Numbers

2.1. Natural Numbers

The positive integers are called natural numbers. They are nothing but abstractions of the quantities that we see around us, i.e. there is a group of ten people, you can have a choice of three different jams for breakfast, or four cars are parking at the side of the street. When humans started to catalog their possessions in the first cities, they used natural numbers to count the objects they possessed. As it would have been too time-consuming to write the complete words for the numbers, humans quickly developed abbreviations for the numbers. Romans used the symbol V for five and the symbol C for hundred. The numerals 1 to 9, that we are using today in the western world, were invented in India and came to Europe via Arabia. It was the Italian mathematician Leonardo Fibonacci who introduced them in the 13th century as he believed that calculations could be much easier performed using Arabian numerals than using Roman numerals.

One reason for this was that Romans simply added up the numerals to describe a number. The different numerals were ordered by size with the largest numerals at the left side. The number one hundred sixty in Roman numerals is: CLX. If Romans wanted to write the number one hundred fifty-nine, they could have written CLVIIII. But this would have been too long. Instead they wrote CLIX, which had to be interpreted as one hundred sixty minus one; for if the smaller numeral came before the larger one, it had to be subtracted. If Romans wanted to write down large numbers, the sequence of numerals quickly became confusing as you also had to consider the position of the numerals to see if you had to add them to the final number – or if you had to subtract them.

The Indians had developed a place-value system that was completely different from the simple sequence of numerals that the Romans used. In the Indian system, the value of a numeral no longer depended on the relative position of the numeral with respect to its neighbors; it only depended on the absolute position within the number.

According to the place-value system that we use today, a number starts with the ones at the right side, followed by the tens on the left, the hundreds on the left again and so on. The number one hundred fifty-nine is thus written as 159 ($1 \cdot 100 + 5 \cdot 10 + 9 \cdot 1$).

If you now wanted to write the number one hundred sixty in the place-value system, you needed a new numeral that the Romans didn't need: The zero, i.e. a numeral that stands for a number which doesn't exist. This sounds like an absurd idea in the first place (having a special symbol for nothing), so only a few cultures invented it. But without a zero, one hundred sixty could only be written as 16, which would be interpreted as sixteen. We are only able to write 160 thanks to the zero.

The place-value system simplifies many calculations. Basic arithmetic operations like addition or subtraction only need to operate on the numerals at the same position (plus considering a carry-over if needed). But it is no longer necessary to consider the whole number while performing calculations.

*

Mathematics doesn't try to establish calculation rules for specific numbers, but general calculation rules that can be applied to all numbers. To achieve this, mathematics uses variables for its laws that can mean any number. Generally, known numbers are represented by small letters from the beginning of the alphabet (a, b, c…), while unknown numbers are represented by small letters from the end of the alphabet (x, y, z). In many cases the multiplication sign is not written, i.e. instead of $a \cdot b$ mathematicians usually write ab.

With these definitions, we can write down some basic calculations rules that are valid for addition and multiplication of natural numbers. These rules are the commutative law, the associative law, and the distributive law.

The commutative law states that you can permute the order of the elements during addition and multiplication:

$$a + b = b + a; ab = ba$$

The associative law states that the addition and multiplication of three numbers will yield the same result independent of the order that you perform the calculation, i.e. it doesn't matter if you first calculate the result for the first two numbers or if you first calculate the results for the last two numbers:

$$a + (b + c) = (a + b) + c; a(bc) = (ab)c$$

The distributive law combines addition and multiplication and states that you can multiply a sum with a number by multiplying each summand with that number and then perform the addition:

$$a(b+c) = ab + ac$$

You can also define an inverse operation to addition and multiplication. The inverse operation for the addition is the subtraction. The inverse operation for the multiplication is the division. The laws just mentioned, however, are no longer valid for subtraction and division. You could try to prove this statement in general terms, but there is a simple rule in mathematics that a law is refuted if you only find one counterexample. So let's assume that the laws would apply. Let's consider the distributive law for the numbers $a = 20, b = 4, c = 2$.

The left side of $a(b - c) = ab - ac$ yields:

$$20/(4-2) = 20/2 = 10$$

The right side yields the result:

$$20/4 - 20/2 = 5 - 10 = -5$$

As you can see, the results are not identical.
In the same way, you can easily show that the associative law is not valid for subtraction:

$$20 - (4-2) \neq (20-4) - 2$$
$$20 - 2 \neq 16 - 2$$
$$18 \neq 14$$

As well as for division:

$$20/(4/2) \neq (20/4)/2$$
$$20/2 \neq 5/2$$
$$10 \neq \frac{5}{2}$$

Exchanging the two numbers $a = 6$ and $b = 3$ in a subtraction yields the result:

$$6 - 3 \neq 3 - 6$$
$$3 \neq -3$$

Similarly, the application of the commutative law to the operation of division results in:

$$6/3 \neq 3/6$$
$$2 \neq \frac{1}{2}$$

When applying the „commutative law" we get some strange results. In the case of subtraction, we are faced with negative numbers. They are not part of the natural numbers, the positive integers that we use for counting. These numbers expand the set of natural numbers. The set of negative and positive numbers is called the set of integers.

When we perform a division, we come to an even stranger result: The result is not even an integer any more, but a number between 0 and 1 that we write in the form of a fraction. Thus, we have to expand the set of integers if we also want to describe these numbers.

2.2. Rational Numbers

The rational numbers were "invented" when people realized that the division of two integers will not always yield an integer. If you divide 8 by 4 the result is 2, if you divide 20 by 5 the result is 4. But if you try to divide 20 by 6, then you have a problem. You can fit the 6 three times in 20, but that leaves you with a rest of 2 that cannot be divided by 6: $\frac{20}{6} = 3 + \text{rest}(2)$.

To solve this problem, fractions were introduced that expand the set of integers to the set of rational numbers. If you now try to divide 20 by 6, you simply write:

$$20/6 = 3 + 2/6 = 3\frac{1}{3}$$

As the fraction $\frac{2}{6}$ contains a 2 as well in the numerator (above the fraction line) as in the denominator (below the fraction line), we can cancel the 2 from the numerator and the denominator and get the fraction $\frac{1}{3}$. The approach to cancel a common multiple in a fraction is a standard approach in mathematics to simplify a fraction.

You can perform your calculation with fractions as easily as with integers. The associative law, the commutative law and the distributive law are also valid for fraction. If you want to add two fraction numbers like $4\frac{2}{3}$ and $5\frac{1}{5}$, you only have to remember that these numbers are nothing but sums and use the distributive law:

$$4\frac{2}{3} + 5\frac{1}{5}$$
$$= 4 + 5 + \frac{2}{3} + \frac{1}{5}$$

One question is now how to add two different fraction like $\frac{2}{3}$ and $\frac{1}{5}$? At first glance, these two numbers seem to be as different as apples and pears. After all, the first number is divided by 3 and the second number by 5. If we would divide both numbers by the same number, then the calculation would be easy: We would simply add both numerators and then divide by the common denominator. Well, and that's what we will do: We find the common denominator and multiply the numerator and the denominator with the number that is needed to create the common denominator. As the numbers 3 and 5 do not have a common multiple, the common denominator for both fractions is simply 15. That means we have to multiply the first fraction by 5 (in the numerator and denominator, i.e. undo a cancellation of 5), and the second fraction by 3. This results in:

$$4\frac{2}{3} + 5\frac{1}{5}$$

$$= 9 + \frac{10}{15} + \frac{3}{15}$$

$$= 9 + \frac{10 + 3}{15}$$

$$= 9\frac{13}{15}$$

And how do we multiply two fractions? Let's write this question in a general form, just like mathematicians like to do it:

$$\frac{a}{b} \cdot \frac{c}{d}$$

$$= (a/b) \cdot (c/d)$$

There is no difference if we divide a number first by a number b and then multiply the result with the number c, or if we multiply first with the number c and then divide by the number b. That means we can write:

$$= a/b \cdot c/d$$
$$= a \cdot c/d/b$$

This means we calculate the product $a \cdot c$ and divide this by d and then by b. But if we are dividing a number by a first number and then by a second number, we could have multiplied both numbers and divided the original number by the product of the two numbers. Therefore, we can write:

$$= (a \cdot c)/(d \cdot b)$$

This expression written as a fraction gives:

$$\frac{a}{b} \cdot \frac{c}{d} = \frac{ac}{bd}$$

That means you multiply two fractions by multiplying the numerators and multiplying the denominators.
The division is the inverse operation of the multiplication. That means we have:

$$\frac{a}{b} / \frac{c}{d} = \frac{a/c}{b/d}$$

This expression is not easy to calculate as a division is usually more complicated than a multiplication. We can modify the expression by expanding the expression with a "one" that doesn't change the value of the expression. In our case, the one will be written as:

$$1 = \frac{cd}{cd}$$

With this approach we get:

$$\frac{a}{b} / \frac{c}{d} = \frac{a/c}{b/d} = \frac{acd/c}{bcd/d}$$

Dividing c by c in the numerator yields a one, just like dividing d by d in the denominator. Therefore we get:

$$\frac{a}{b} / \frac{c}{d} = \frac{ad}{bc}$$

We divide two fractions by multiplying the first fraction with the inverse of the second fraction.

In ancient times, people believed that the rational numbers would describe all numbers that exist. After all, you could perform the basic calculation of addition, multiplication, subtraction and division and you would come up with a rational number again.

However, already the ancient Greeks discovered that the rational numbers could not be the whole story. They discovered this by taking a closer look at the Pythagorean Theorem.

2.3. Real Numbers

The Pythagorean Theorem was proved in the 6th century B.C. by the Greek philosopher Pythagoras of Samos. The relationship between the sides of a right-angled triangle mentioned in the theorem was already known by the ancient Egyptians and Babylonians. But there is no record that they proved the correctness of the relationship for all triangles in general.

Figure 1 shows a right-angled triangle. The sides adjacent to the right angle are called cathetus or legs, the side opposite to the right-angle is called hypotenuse.

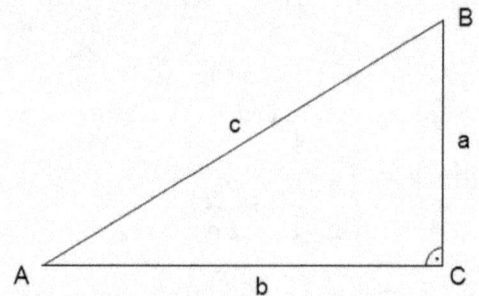

Figure 1: A right-angled triangle. The sides adjacent to the right angle are called legs, the side opposite to the right angle is the hypotenuse.

According to the Pythagorean Theorem, the following relationship is valid in a right-angled triangle:

$$a^2 + b^2 = c^2$$

The superscripted 2 is called an exponent. The exponent 2 means that the variable a is multiplied twice with itself

($a^2 = a \cdot a$). An exponent 5 would mean that the variable a is multiplied five times with itself: $a^5 = a \cdot a \cdot a \cdot a \cdot a$.
The inverse statement is also true: If we have a triangle where the relation $a^2 + b^2 = c^2$ is valid, then this triangle has to be a right-angled triangle, with the right angle opposite to the side c.

You can prove the Pythagorean Theorem in many ways. One simple way is shown in figure 2. For this proof we take four triangles with the sides a, b, and c, and construct an outer square with the side $a + b$ and an inner square with the side c.

Figure 2: One possibility to prove the Pythagorean Theorem: Construct an outer square made of four triangles. The length of the sides of this square is a+b. The inner square has sides with the length of c. Calculating its area yields: a² + b² = c²

The outer square has the area $(a + b)^2$. The inner square has the area c^2. To get from the outer square to the inner square, you have to subtract the area of the four triangles. The area of a triangle is $\frac{1}{2}ab$ (half of the area of the respective rectangle). This means that we have:

$$c^2 = (a+b)^2 - 4 \cdot \frac{1}{2}ab$$
$$= a^2 + 2ab + b^2 - 2ab$$
$$= a^2 + b^2$$

which is exactly the Pythagorean Theorem (mathematicians like to decorate their proofs with the abbreviation "q.e.d." which stands for "quod erat demonstrandum" and means "which had to be demonstrated").

It is now easy to show that the inverse theorem is also valid, i.e. that a triangle where the relation $a^2 + b^2 = c^2$ is valid has to be a right-angled triangle. The argument is as follows:

Assume that you have a triangle with the sides a, b, and c, and that the relation $a^2 + b^2 = c^2$ is valid for this triangle. Now we want to show that the angle between the sides a and b is a right angle. To do this, we construct a second triangle with the sides a and b, which includes a right angle between these sides. According to the Pythagorean Theorem, the hypotenuse has the length: $c = \sqrt{a^2 + b^2}$. This means that all sides of the second triangle have the same length as the first triangle. They are, as the mathematicians say, congruent. When they have the same sides, they also have to have the same angles. As the angle between a and b in the second triangle is a right angle, the angle between a and b in the first triangle had been a right angle as well.

In this proof we have used the inverse of raising a number to a power: The root. When you perform an exponentiation, you multiply a number as often with itself as the exponent tells you. If you extract a root, you are looking for the number that yields a given result if multiplied with itself. Thus, the root of 9 is 3 (you write this as $\sqrt{9} = 3$), because three multiplied with three yields nine. You can also define roots for higher exponents. So you have $\sqrt[3]{81} = 3$, as the three multiplied with itself three times yields 81.

In general, the extraction of the root doesn't yield natural numbers but any numbers. Unfortunately, there is no easy way to extract a root for any number. Using the logarithm (we will talk about this later) made it possible to approximate the solution fairly well. Today, the availability of pocket calculators solves this problem anyway.

The ancient Greeks took the Pythagorean Theorem and thought they would be solving a very easy case when they assumed that the legs of the triangle had a length of 1. Now what is the length of the hypotenuse?

According to the theorem we have:

$$c^2 = 1^2 + 1^2$$
$$c^2 = 2$$
$$c = \sqrt{2}$$

To be precise, the Greeks were not the first to find the stranger number $\sqrt{2}$. It was already known to the ancient Babylonians who tried to write this number as a fraction. The best approximation they came up with was the fraction $\frac{577}{408}$, which at least reproduces the five positions after the decimal point correctly.

The Greeks spent a lot of time trying to figure out the rational number that would be identical to the root of 2. It was a Pythagorean who showed that this is not possible. The proof for this is rather simple:

1. Let us assume that $\sqrt{2}$ is a rational number that can be written as a fraction with a numerator a and a denominator b, so that we have: $\frac{a}{b} = \sqrt{2}$.
2. The numbers a and b have been chosen so that they do not have a common multiple.

3. The assumption $\frac{a}{b} = \sqrt{2}$ results in $\frac{a^2}{b^2} = 2$ and finally $a^2 = 2b^2$.
4. This means that a^2 can be divided by 2, i.e. it is an even number.
5. Therefore a is an even number as well (as a was multiplied with itself, a has to contain a 2 if the square of a contains a 2).
6. As a is an even number, we can find a number k for which the following relation is valid: $a = 2k$.
7. We are now replacing $a = 2k$ in the equation $a^2 = 2b^2$ from step 3. This results in: $(2k)^2 = 2b^2$. This expression can be simplified to $b^2 = 2k^2$.
8. This means that b also has to be an even number.
9. We get the result that a and b have to be even numbers, i.e. they contain the common multiple 2 in contradiction to our assumption that they do not have a common multiple.
10. The assumption that $\sqrt{2}$ can be written as a fraction $\frac{a}{b}$ leads directly to a contradiction, i.e. the assumption has to be wrong.
11. q.e.d.

In ancient times, the Greeks believed that all numbers could be written as fraction, as rational numbers. But the number $\sqrt{2}$ showed them that this is not the case. Quite obviously there are some numbers that cannot be written as a fraction. You can approximate the number $\sqrt{2}$ with the number 1.41421356…, but you will never be able to describe it exactly in this way. If we want to take care of all numbers, we have to expand the set of rational numbers to the set of real numbers.

Another example of a real number is a number that is simply described as pi (π). Already in ancient times the people knew that the area of a circle is proportional to the square of the radius, while the circumference is proportional to the diameter (i.e. twice the radius). The proportionality factor seemed to be identical in both cases. It was simply called π and we got the following two formulas for a circle:

Circumference: $2\pi r$

Area: πr^2

For several centuries, mathematicians tried to find a simple expression for the value of π. In the 3rd century B.C. Archimedes came up with this estimation:

$$3{,}1408450 \approx 3 + \frac{10}{71} < \pi < 3 + \frac{10}{70} \approx 3{,}1428571$$

Today's value for π is: $\pi = 3{,}141592654...$ However, it was only in the 18th century that the mathematician Johann Heinrich Lambert proved that π is not a rational number and cannot be written as a fraction.

Now that we have the set of real numbers, all possible numbers seem to be covered.

But as we just started to extract the root of numbers – what is the root of -1?

2.4. Complex Numbers

One area of mathematics deals with systems of equations. Systems of equations are equations with one or more unknowns that have to fulfill all given equations at the same time. The simplest equation is the linear equation with just one unknown:

$$ax + b = 0$$

This equation is called linear as the unknown appears with the exponent one: x^1, which is not written for simplicity reasons. The solution can be found easily:

$$x = -\frac{b}{a}$$

The case is somewhat more complicated if we have a quadratic equation, i.e. an equation where the unknown has an exponent of 2:

$$ax^2 + bx + c = 0$$

To solve this equation, we divide the whole equation by a (which gives us a 1 as a factor before the x^2) and replace $\frac{b}{a}$ by p and $\frac{c}{a}$ by q. With this we get the so-called normal form of the equation:

$$x^2 + px + q = 0$$

This equation is solved with a small trick. This trick consists in adding the same term on both sides which doesn't have an influence on the sought value for x. Before we do this, we bring q to the other side:

$$x^2 + px = -q$$

Now we add $\left(\frac{p}{2}\right)^2$ on both sides and get:

$$x^2 + px + \left(\frac{p}{2}\right)^2 = \left(\frac{p}{2}\right)^2 - q$$

We can simplify the left side to:

$$\left(x + \frac{p}{2}\right)^2 = \left(\frac{p}{2}\right)^2 - q$$

Now we extract the root on both sides and get:

$$x + \frac{p}{2} = \pm\sqrt{\left(\frac{p}{2}\right)^2 - q}$$

The root of a number can be positive or negative, as squaring a negative number will also result in a positive number. Extracting the root is therefore not non-ambiguous.

This finally gives us a value for the unknown x.

$$x = -\frac{p}{2} \pm \sqrt{\left(\frac{p}{2}\right)^2 - q}$$

If you take the quadratic equation $x^2 + 6x + 90 = 0$, the solution is:

$$x = -\frac{6}{2} \pm \sqrt{\left(\frac{6}{2}\right)^2 - 90}$$

$$x = -3 \pm \sqrt{9 - 90}$$

$$x = -3 \pm \sqrt{-81}$$

Now, which number multiplied with itself will yield -81? Multiplying a positive number with itself will result in a positive number. Multiplying a negative number with itself will result in a positive number as well. There is simply no real number that will yield a negative number if multiplied with itself, i.e. we have to expand the set of numbers again.

To simplify the attempt, mathematicians single out the troublemaker, i.e. the minus (one) and write the $\sqrt{-81}$ as $\sqrt{-1} \cdot \sqrt{81}$. The root of 81 is simply 9. The $\sqrt{-1}$ is designated as *i*. With this nomenclature, the solution of the quadratic equation is:

$$x = -3 \pm 9i$$

This number is called a complex number. The number -3 is the real part and the number $9i$ is the imaginary part of the complex number. It was only in the 17th century when mathematicians started to take a closer look at the roots of negative numbers. And it was during this time that they started to call the "normal" numbers that were known up to then "real numbers" while the root of a negative number was called an "imaginary number".

There is no way that we could attach a meaning to imaginary numbers in our real world, as you can attach units like meter or liter to a real number. That's why these numbers were called imaginary.

Mathematics, however, is no natural science that deals with objects of the real world. Mathematicians live in the world of their own objects, so they fruitfully work with imaginary numbers as if they were the real thing.

To tell the truth, dealing with complex numbers is not just an exercise of the mind. Physicians can simplify some calculations, especially if they deal with trigonometric functions, when they use a "detour" through the field of complex numbers. But we will talk more about this later.

Even though mathematicians work with abstract objects that don't exist in the real world, they like to use pictures to visualize what they are dealing with, just as they like to draw a triangle, even though its sides have a certain width if drawn. So they also would like to visualize complex numbers.

The real numbers are shown on a number ray. This number ray goes in both directions, away from its point of origin, the zero, and it allows placing any given real number. Complex numbers, however, do not fit on a number ray. Therefore, mathematicians consider the imaginary part of the complex number to be an additional coordinate. Complex numbers can be visualized on a complex plane, as shown in figure 3.

Such a coordinate system where a point is non-ambiguously defined by its coordinates is called a Cartesian coordinate system. René Descartes was a French mathematician and philosopher who lived in the 17th century. Such a coordinate systems makes it possible to describe an object by writing down the coordinates of its sides and corners, thus allowing us to treat geometric objects analytically, i.e. to calculated with them as if they were numbers. Descartes was one of the founders of analytical geometry, although he likely didn't come up with the coordinate system that bears his name. He was, however, the first one who called the components of a complex number "real" and "imaginary" part.

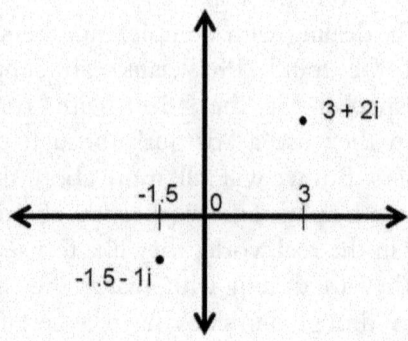

Figure 3: Real numbers can be placed on a number ray. The complex numbers form the complex plane.

The calculation rules for complex numbers are fairly easy. An addition and a subtraction is performed on the respective real and imaginary parts of the complex numbers, as if you were dealing with two different physical units. This means we have:

$$(a + bi) + (c + di) = (a + c) + (b + d)i$$

and

$$(a + bi) - (c + di) = (a - c) + (b - d)i$$

The multiplication is a simple calculation:

$$(a + bi) \cdot (c + di)$$

$$= ac + bci + adi + bdi^2$$

$$= (ac - bd) + (ad + bc)i$$

as per definition we have $i^2 = -1$.

But how do you divide a complex number by another complex number? To have a complex number of the form $a + bi$ after the division, you need to get rid of the complex number in the denominator. This is done again by expanding the fraction with a one, this time in the form of $\frac{(c-di)}{(c-di)}$. With this approach we get:

$$\frac{a + bi}{c + di}$$

$$= \frac{(a + bi)(c - di)}{(c + di)(c - di)}$$

$$= \frac{ac + bd}{c^2 + d^2} + \frac{bc - ad}{c^2 + d^2}i$$

We started with natural numbers and had to expand the set of number until we came up with the complex numbers. Is this all, or will there be more? After all, complex numbers describe a plane, could there be something like a three summand complex number that describes a volume?

In principle, this could be possible. But mathematicians prefer that their objects follow certain rules of calculation like the commutative law. This is not possible for three dimensions. Mathematicians can only describe a meaningful "number object" in four dimensions that they call quaternions. They have the form $a + bi + cj + dk$, with the real part a and the imaginary part $bi + cj + dk$. These numbers, however, are not used in natural science, they are pure mathematical objects, so we will not look at them further.

2.5. The Rule of Three

Mathematicians like to prove things, like the Pythagorean Theorem or the fact that the root of 2 is not a rational number. But mathematics can also be used in daily life.
Let's take the famous promotional campaign "Today, you don't have to pay the value-added tax". There are several numbers for value-added tax in different countries, in Germany the number is 19%. If a product costs 200 Euros, how much money can you save if you don't have to pay the VAT?
This is the kind of tasks we never liked at school. But let's try to solve the question. The 19% means nothing else but 19 from hundred (percent is derive from Latin and means literally "from hundred").
So 19% is nothing else but $\frac{19}{100} = 0.19$.
If the price is 200 Euros, then 19% are $200 \cdot 0.19 = 38$. So you can save 38 Euros, right?
You know this approach from school: If this question is asked, the answer has to be no. A percentage always needs a point of reference. What is the point of reference for the VAT? Is it definitely not the final price, it is not the 200 Euros you have to pay. The point of reference is the net price that will give 200 Euros if you add the VAT. The Net price plus the VAT result in the gross price that you will pay at the cash register. The net price is this 100%, the VAT in our example is 19%, and the gross price is therefore 119%. Let's summarize what we have so far:

A saving of x Euros is 19%.
The end price of 200 Euros is 119%

We are looking for the unknown saving of x Euros thanks to the promotional campaign. How can we calculate this saving?

The approach is to find out how many Euros we save with each percent-point. To get this number, we divide the 200 Euros by 119%. This gives us a price per percent point of about 1.68 Euros. If we multiply this with the 19% that we save, we find that the VAT actually was roughly 31.92 Euros. Written as a compact formula, we have:

$$x = \frac{200\ \text{€}}{119\%} \cdot 19\% = 31.92\ \text{€}$$

The net price was 200 € − 31.92 € = 168.08 €. The test shows that a VAT of 19% added to a net price of 168.08 € really results in an end price of 200 €.

Let's now write the formula in a more symmetric way:

$$\frac{200\ \text{€}}{119\%} = \frac{x}{19\%}$$

Or in a more general form:

$$\frac{a}{b} = \frac{x}{c}$$

The proportion between the given numbers for the end price and the partly unknown numbers for the saving are equal. If the end price rises, the saving also rises. The more we have of a, the more we will have of x. We have a proportional correlation.

To solve such a proportional correlation, we can summarize the given and unknown numbers in a table. Values of the same kind are in the same column. Traditionally, the unknown number is at the bottom right. This gives the picture shown in table 1.

1. Value	2. Value
b	a
c	x

Table 1: The rule of three.

To calculate the value for one percent, we first divide a by b, then we multiply the result by c to get x. To memorize this approach, we simply remember that we multiply the diagonal $a - c$ and divide the result by b to get x. This calculation is called rule of three as three known numbers help us to calculate an unknown fourth number.

One application for this rule of three can be the comparison of prices. One big package of detergent of 1kg costs 7.90 Euros, a smaller package of 300g only costs 2.49 Euros. Which offer gives you more detergent for the same amount of money?

Let's take the 1kg-package as a starting point. What would be the price of the 300g package if the price-relation would be the same as for the 1kg package? If we have to pay more for the actual 300g-package than this theoretical price, then you get more detergent for your money if you buy the 1kg package. If the price of the 300g-package is lower than this theoretical price, then you better buy the smaller package.

Amount	Price
1000g	7.90€
300g	x

Table 2: Example for the rule of three.

The correlation is summarized in table 2. This brings us to the calculation:

$$x = \frac{300g \cdot 7.90\,€}{1000g} = 2.37\,€$$

So we would have to pay less for 300g taken from the 1kg-package than we actually have to pay for the 300g-package. The larger package gives you more detergent for your money.

Now, if the there is a proportional correlation, we might assume that there is also something like an inversely proportional correlation. And in fact, there is.

Let's consider the example of a treasure that can be shared equally among some sailors. If there are ten sailors, each one will get 50 gold coins from the treasure. How many gold coins would each sailor get if there were 20 sailors?

We can easily see the difference between an inversely proportional and a proportional correlation: In the case of a proportional correlation the respective amounts grew in the same proportion. A higher percentage meant a higher price. In the case of an inversely proportional correlation, the situation is completely different: If we have to share the treasure between more sailors, each one will get less, as the number of gold coins in the treasure is fixed. In such a case, we start our calculation with finding out how much there is that can be shared.

We said that ten sailors could get 50 gold coins each, that means we have $10 \cdot 50 = 500$ gold coins to share. If we share them between 20 sailors, then each sailor will only get $500 : 20 = 25$ gold coins. Written in the form of a compact formula, we have:

$$x = \frac{10 \cdot 50}{20} \text{ gold coins}$$

In a more symmetrical form:

50 gold coins $\cdot$ 10 sailors $= x \cdot$ 20 sailors

Or in a general form with variables:

$$a \cdot b = x \cdot c$$

Where we had a division in the case of a proportional correlation, we have a multiplication in the case of an inversely proportional correlation, because in the case of a proportional correlation we first want to find out the price of one unit, in the case of an inversely proportional correlation we first want to find out the complete amount there is.

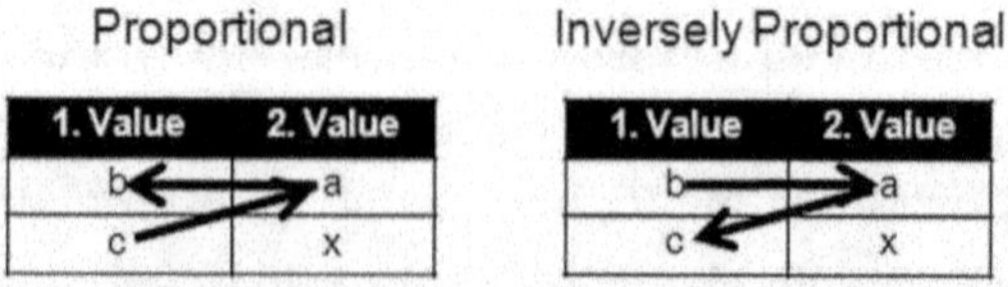

Proportional		Inversely Proportional	
1. Value	2. Value	1. Value	2. Value
b	a	b	a
c	x	c	x

Figure 4: Calculation rule for the proportional and the inversely proportional correlation: For the proportional correlation we first multiply a with c and then divide by b, for the inversely proportional correlation we multiply a with b an then divide by c.

With a similar table as table one, we can easily calculate an inversely proportional correlation. Only in this case you first multiply the numbers in the first line and then divide by c. Figure 4 summarizes the approaches for the proportional and inversely proportional correlation.

3. Functions

3.1. Mappings

If you heat up a piece of metal it will expand. But what is the expansion of the piece of metal for a certain temperature? If you exert a certain force on an object, it will change its movement. But how big is the change of movement, the acceleration, for a given force?

We quite often have the case that the change of one physical variable influences another physical variable. Mathematicians took this observation as an input to develop a new mathematical field.

You are changing a variable like force or temperature and this changes another variable like acceleration or the size of a body. If this is the case, mathematicians say that you map one set of variables onto another set of variables. The calculation rule that defines the relation between these two sets is called a function. All a function does is to map one set of numbers onto another set of numbers.

The set of inputs and the set of outputs are generally real numbers (but they can also be complex numbers or even functions that are mapped onto other functions). The probably simplest relation is the identical relation:

$$y = x$$

In this case, each number is related to itself. Generally, the set of inputs is designated with the variable x and the set of outputs with the variable y. If you set $x = 2$, then you get a 2 again (in our example). If you interpret the x and y as coordinates in a plane, and if you depict all y-values that you generate from the x-numbers, then you get the graph of the function. In our case, the graph is simply a diagonal through the point of origin, as can be seen in figure 5 (traditionally, the x-values are plotted along the horizontal axis, the y-values along the vertical axis).

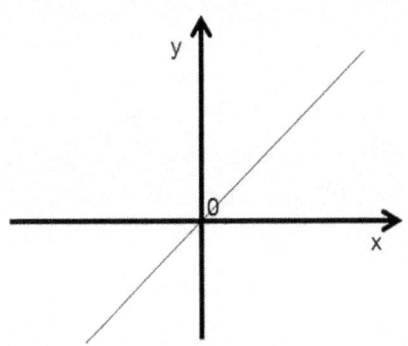

Figure 5: The function y = x is a diagonal through the point of origin.

It may now happen that a function not only depends on one variable, but on several variables. The mathematician Leonard Euler thus introduced another form to designate the relation between the variables. He proposed to call the function simply f and add the variable(s) in brackets. In the case of the diagonal, this is:

$$f(x) = x$$

If the function were depending on the variables x and y, then we would write for example:

$$f(x, y) = ax + by$$

Using this nomenclature, we have no problem to identify which letters are used for a variable of the function and which are constant.

A quadratic function of the form

$$f(x) = ax^2 + b$$

describes a parable. The coefficient a shows, how broad the parable opens up (the bigger a is, the narrower the parable becomes, as you can see if you calculate some number pairs $(f(x), x)$ for different values of a). The summand b shows us where the parable meets the y-axis. Figure 6 shows the parable $f(x) = 2x^2 + 1$.

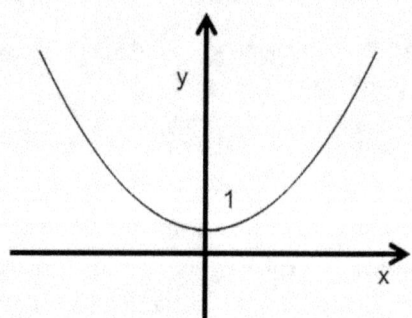

Figure 6: The function f(x) = 2x² + 1 is a parable through the point y=1.

The amount of possible functions is endless, as you can combine all possible combinations of x and its exponents with all possible constants. But there are also some function where you do not raise x to a certain power, multiple or add it. And some of these functions are of special interest in physics. For instance, there are functions where you do not raise x to a certain power, but where x is the exponent. These functions are called exponential functions. The inverse of an exponential function is the logarithm.

3.2. Exponential Function and Logarithm

If we raise a number to a certain power, the exponent tells us how often we have to multiply the number with itself: $a^2 = a \cdot a$.
If the number a is a variable and you can take any numbers for $a = x$ then we have a parable: $f(x) = x^2$.
Such a function is called a power function: The exponent is fixed; only the basis will be changed. An exponential function, on the other hand, keeps the basis fixed and changes the exponent. The general form of an exponential function is:

$$f(x) = a^x$$

In the old times, before pocket calculators became available, you had to look into thick tables to figure out the values of exponential functions with non-integer exponents.
Some simple rules apply for exponential function:

$$a^0 = 1$$
$$a^1 = a$$

$$\frac{1}{a} = a^{-1}$$

$$\frac{1}{a^x} = a^{-x}$$

$$\sqrt[q]{a^p} = a^{\frac{p}{q}}$$

Furthermore, we have for the multiplication:

$$a^{x+y} = a^x \cdot a^y$$

$$a^{x \cdot y} = (a^x)^y$$

$$a^x \cdot b^x = (a \cdot b)^x$$

For increasing values of x, the value of an exponential function increases very quickly. Let's take the basis of 10, then $x = 1$ gives the value 10, $x = 2$ yields 100, and $x = 3$ gives the results 1000. Figure 7 shows the graph of the exponential function $f(x) = 10^x$.

For increasing x-values, the values of the exponential function quickly increase to infinity, while the exponential function goes to zero for negative x-values – but it never reaches zero (as we have $10^{-1} = \frac{1}{10} = 0.1$, $10^{-2} = 0.01$ and so on). The exponential function does not yield negative values. Its set of outputs is always positive.

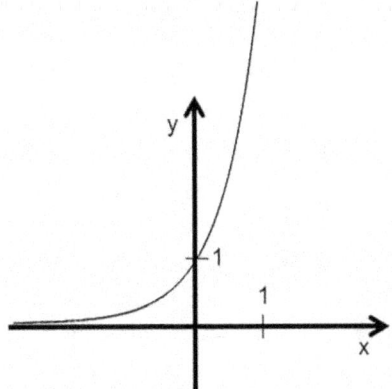

Figure 7: The function $f(x) = 10^x$.

The inverse operation for the addition is the subtraction. Accordingly, there is also an inverse operation for the exponential function. It is called logarithm.

If you have an exponential function, then you know the Basis a and wonder, which number y you get if you raise a to the power of x: $y = a^x$.

The inverse function now asks to what power you need to raise the basis a to get the result y. The logarithm (written as log) is then: $\log_a y = x$. Now the exponent is the result of the calculation. The basis a is noted as a subscript. If your basis is 10, then we usually write lg instead of $\log_{10}$. With the known notation (the unknown variable is the x) the logarithm function can be written as:

$$f(x) = \log_a x$$

The logarithm is only defined for input values of $x > 0$, because the exponential function only provides positive output values – and the logarithm, as the inverse of the exponential function, can only be "fed" with those numbers that the exponential function provides.

If we have $x = 1$ then the logarithm yields 0 (after all we have $a^0 = 1$). For values smaller than $x = 1$ the logarithm approaches $-\infty$ (the horizontal 8 stands for the number "infinity"), because for negative inputs the exponential function yields smaller and smaller outputs that approach the zero.

For values bigger than 1 the logarithm increases as well. But while the exponential function literally "explodes", the logarithm grows slowly as you have to cover one order of magnitude to increase the logarithm by one unit (in the case of the basis 10). The typical graph of a logarithm is shown in figure 8.

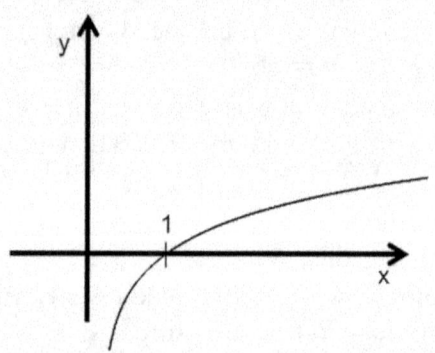

Figure 8: Typical behavior of a logarithm $f(x) = \log_a x$.

There are some nice calculation rules for the logarithm that explain while the logarithm was widely used before the introduction of a pocket calculator:

$$\log_a(x \cdot y) = \log_a x + \log_a y$$

$$\log_a \frac{x}{y} = \log_a x - \log_a y$$

$$\log_a(x^r) = r \cdot \log_a x$$

These rules can be used to replace multiplication and division by addition and subtraction. This provides a significant simplification, especially if you are dealing with big numbers. If you wanted to solve the division **589345789 : 446**, you needed some patience – or you could use the logarithm:

$$\lg\left(\frac{589345789}{446}\right) = \lg(589345789) - \lg(446)$$

(here we are using the logarithm to the basis of ten). With this approach you can replace the division by a subtraction. All you needed was the logarithm of different numbers. But they were available in large tables and cold simply be looked up. So you got

$$\lg\left(\frac{589345789}{446}\right) = \lg(589345789) - \lg(446)$$

$$= 8.77037 - 2.64933$$

$$= 6.12104$$

This result is the exponent to the basis of ten. The calculation yields (again you just looked it up in the tables): 1321417.336. The pocket calculator gives a result of 1321403.114 which is quite close to our result. And in the times before the pocket calculator, this was better than nothing.

The logarithm was especially useful when it came to extract the xth root auf a number. If you wanted to calculate the 4th root of 1219, you simply used the third calculation rule:

$$\lg (\sqrt[4]{1219} = \lg \left(1219^{\frac{1}{4}}\right) = \frac{1}{4}\lg (1219)$$

The logarithm to the basis of ten of the number 1291 could be looked up, and you get 3.0860. Divide it by four and you have 0.7715. This is the logarithm of the final result. We invert again the logarithm and get 5.91, which is the 4th root of 1219.

Even though the logarithm seems to be quite complicated, it helped to simplify calculations significantly, especially in times before the pocket calculator. Today, we find the logarithm tables mainly in second hand bookshops.

But this doesn't mean that we don't have any use for exponential function and the logarithm now that pocket calculators took over most of the work. Scientist and engineers still use these functions to a great extend. However, they rarely use the logarithm to the basis of 10, but usually prefer another basis that was simply called e by Leonard Euler. The number e is a rather strange number, but the logarithm with this basis is nevertheless called the natural logarithm.

The number e is irrational and has the approximate value of 2.718281828459045235…. Euler didn't find this number. It originally appeared in the calculation of interest. But Euler was one of the first who discovered its importance for mathematics and physics.

The number e is irrational which means that it cannot be written as a fraction. And unlike $\sqrt{2}$ the number e is even transcendental which means that it cannot be derived as the solution of an algebraic equation (as it is the case for $\sqrt{2}$ and the Pythagorean Theorem). So there is no way to derive e in a simple way, you can only try to approximate it. The number e was detected rather accidentally in the 17th century when Jacob Bernoulli took a closer look at the calculation of interest.

Let's say that you start with an amount of money K_0. After one year you add the interest and get the new amount of money K. Let's z be the interest rate. Then we can calculate the new amount of money by multiplying the starting amount K_0 with the interest rate and adding this to the starting amount:

$$K = K_0 + z \cdot K_0$$

This can also be written as

$$K = K_0(1+z)$$

In the second year, your amount of money increases to

$$K_2 = K(1+z)$$
$$= K_0(1+z)(1+z)$$
$$= K_0(1+z)^2$$

In general we have the formula that the amount of money after n years is:

$$K_n = K_0(1+z)^n$$

This is the well-known compound interest formula.
Mathematicians like to simplify their calculations. So let's assume we have $K_0 = 1$ and $z = 1$ (even though an interest rate of 100% is not very likely in the real world). With this we get after one year:

$$K_1 = (1 + 1)^1$$

This result is trivial. But it shows an inner beauty when we calculated the interests for shorter periods of time. Let's assume we don't want to cash in our interest once a year but every six month. The interest rate in our example for half a year is $\frac{1}{2}$. As we get it twice a year, we have to raise the expression in the brackets to the power of two:

$$K_2 = \left(1 + \frac{1}{2}\right)^2 = 2.25$$

Now we would like to get our interests weekly. With $n = 52$ weeks per year we get:

$$K_{52} = \left(1 + \frac{1}{52}\right)^{52} = 2.69\ldots$$

If we get our interests daily, we have:

$$K_{365} = \left(1 + \frac{1}{365}\right)^{365} = 2.714567\ldots$$

If we get our interests instantly, i.e. n goes towards infinity, we get the number $e = 2.7182818\ldots$ (So you see: The

difference between getting the interest daily and instantly is only minor. You always have to know when to stop).
When mathematicians say that a variable or a function should go towards a certain value, they usually write:

$$\lim_{n \to \infty} n$$

This means that the variable n goes towards the limit infinity (lim stands for the Latin word limes which mean limit or boundary). Using this expression, we can write the number e as

$$e = \lim_{n \to \infty} \left(1 + \frac{1}{n}\right)^n$$

The inverse of the exponential function to the basis e

$$f(x) = e^x$$

is the natural logarithm

$$f(x) = \ln x$$

The same calculation rules apply for the exponential function to the basis of e (sometimes also written as $f(x) = exp(x)$) and its inverse as for all other exponential functions and logarithms.
The e-function seems to be strange as its basis is a rather weird number; it is not a fraction and not even an integer. But the e-function has the big advantage that its derivative is an e-function again. We will learn more about derivatives later. For the moment we can describe the derivative as the

slope of a function. The value of the e-function for a certain x-value thus also describes its slope at that point.

This is just one advantage and one reason for the popularity of the e-function in science. Another reason is a formula that was discovered by Leonard Euler. This formula relates the e-function to the trigonometric functions sine and cosine. These functions are used in physics to describe oscillations and waves. If you consider that quantum mechanics mainly deals with describing the behavior of material waves, you might imagine that the e-function is one of the most used functions in physics.

But before we can derive Euler's formula, the relationship between the e-function and the sine and cosine, we have to discuss the trigonometric functions first.

3.3. Trigonometric Functions

Let us go back to the right-angled triangle. The Pythagorean Theorem provides a simple way to calculate the lengths of its sides. Using trigonometric functions, we can also calculate the size of the angles based on the length of the sides. Figure 9 shows this for the angle α. The trigonometric functions of the sine, cosine and tangent are defined as:

$$\sin \alpha = \frac{\text{opposite side}}{\text{hypotenuse}} = \frac{a}{c}$$

$$\cos \alpha = \frac{\text{adjacent side}}{\text{hypotenuse}} = \frac{b}{c}$$

$$\tan \alpha = \frac{\text{opposite side}}{\text{adjacent side}} = \frac{a}{b}$$

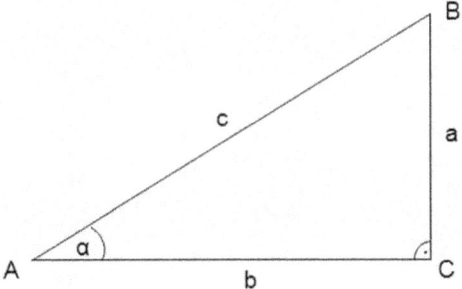

Figure 9: The sine of the angle α is the opposite divided by the hypotenuse, the cosine is the adjacent divided by the hypotenuse and the tangent is the opposite divided by the adjacent.

With this, we are able to calculate all angles when the sides are given (the angle β at the point B can be calculated as $\beta = 180° - 90° - \alpha$).

The inverse function to the sine, cosine and tangent are the arcsine (arcsin or $\sin^{-1}$), the arccosine (arccos or $\cos^{-1}$) and the arctangent (arctan or $\tan^{-1}$), respectively. With these functions we can calculate the angles and sides of a triangle if only two values (side or angle) are available.

The values for the trigonometric functions could be found in large table (like for the logarithm), but today you simply calculate them with the pocket calculator.

The definition of the trigonometric function at the right-angled triangle is very descriptive, but it has the disadvantage that this definition can only be used for angles between $0 < \alpha < 90°$ (because for $\alpha = 0°$ and $\alpha = 90°$ there is no triangle left). This set of inputs of the trigonometric functions is even more restricted than the set of inputs of the logarithm (this one at least includes all positive numbers). This restriction was considered to be unsatisfactory, so mathematicians came up with a new definition for the trigonometric functions.

This definition uses a unit circle, i.e. a circle with the radius $r = 1$. In a unit circle, an angle can have values of 360° and even more if you just walk around the circle several times.

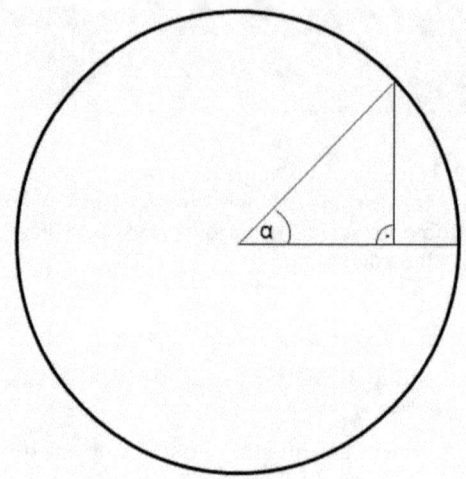

Figure 10: The definition of the radian: The whole angle of a circle is 360°. This is equal to 2π at the unit Circle. An angle of α=45° cuts a direct edge of π/4 out of the circle.

The unit of angles in a unit circle is no longer the degree but the radian. The radian describes the length of the circumference that an angle cuts out of the circle, as figure 10 shows.

The complete circumference is 360°. For a unit circle the circumference is $2\pi r^2 = 2\pi$ (as the unit circle has a radius of $r = 1$). Half a circle thus has a radian of π (an angle of

180°), and a right angle has a radian of $\frac{\pi}{2}$. Larger angles are described by higher multiples of π.

The use of the unit circle now allows for a definition of the trigonometric functions with arbitrarily large input values. As the hypotenuse of the inner triangle is equal to the radius (which is just $r = 1$), the definition of the sine and cosine is fairly easy:

$$\sin \alpha = \text{opposite}$$
$$\cos \alpha = \text{adjacent}$$

You can look at the sine and cosine as coordinates of a point P on the circumference, if the point of origin is in the center of the circle, as figure 11 shows.

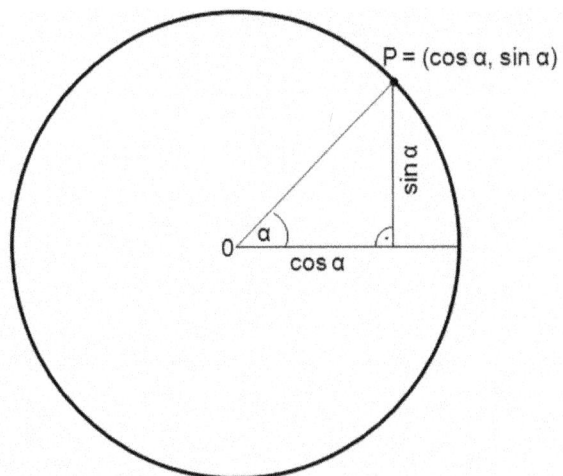

Figure 11: The definition of sine and cosine at the unit circle. They can be considered to be the coordinates of the point P on the circumference of the circle if the point of origin is at the center of the circle.

The tangent is defined as the opposite side divided by the adjacent side, i.e. we have:

$$\tan \alpha = \frac{\sin \alpha}{\cos \alpha}$$

Now we have spent quite some time defining the trigonometric functions. But what do these functions actually look like? Let's derive the graph for the sine function. We use the unit circle as shown in figure 12.

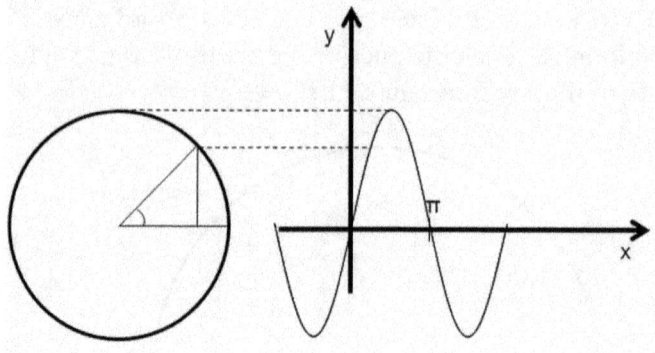

Figure 12: The function sin x.

If the angle is zero, than we obviously also have $\sin 0 = 0$. The function thus starts at the point of origin. If the angle becomes bigger, the sine also increases. At the angle of $\frac{\pi}{4}$ (45°) the sine has a value of $\frac{\sqrt{2}}{2}$ ($\approx 0{,}71$) as both legs of the right-angled triangle have the same length. For greater angles up to the right angle $\left(\frac{\pi}{2}\right)$ the growth of the sine is a little smaller as the circle levels off. The sine has reached its maximum value at $y = 1$ (after all, we are looking at the unit circle), and it becomes smaller for increasing radians.

For values bigger than π the sine is even negative, as the angle is now moving through the lower part of the circle.

For angles between 2π and 3π, we have a positive sine again, as we are now moving through the upper half of the circle, and so on.

Based on this derivation, we can easily see what the cosine should look like: If the sine is minimal, the cosine is at its maximum and if the sine is at its maximum, the cosine is zero. The cosine function thus starts at the coordinate (0, 1). Its graph is shown in figure 13.

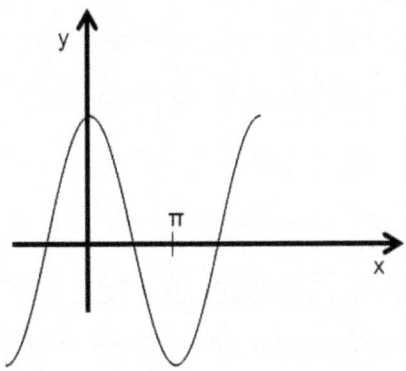

Figure 13: The function cos x.

Sine and cosine are thus shifted by $\frac{\pi}{2}$ and we have:

$$\sin x = \cos(x - \frac{\pi}{2})$$

And, of course, the Pythagorean Theorem is still valid:

$$\sin^2 x + \cos^2 x = 1$$

Mathematicians usually write the exponent of a trigonometric function directly behind the name of the function, i.e. $\sin^2 x$ instead of $(\sin x)^2$.

The tangent is defined as the quotient of sine and cosine. The cosine is zero for the values of $\frac{\pi}{2}, \frac{3}{2}\pi$ and so on, i.e. the tangent will go towards infinity for these values (you can't divide a number y zero, i.e. the tangent for this numbers is not defined, but it becomes pretty large for values close by). Furthermore, the sine is zero for $0, \pi, 2\pi$ and so on, i.e. the tangent has to be zero here as well. This leads us to the graph of the tangent as shown on figure 14.

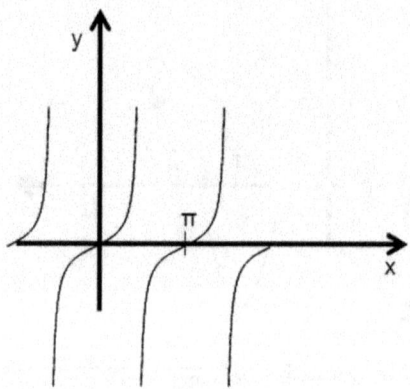

Figure 14: The function tan x.

If you are calculating the values of trigonometric functions with your pocket calculator, you have to pay attention if it is set to radian or degree – the results will be completely different for a given number.

The inverted trigonometric functions can easily be derived from the trigonometric functions. You just have to consider that the set of inputs for the arcsine and the arccosine can only be taken from the interval $[-1, 1]$, as this is the possible outputs of the sine and cosine. The output values for the arcsine and the arccosine, however, can be infinitely large, as the input interval for the sine and cosine was not limited. The arcsine will be zero for $x = 0$, while the arccosine will be $\frac{\pi}{2}$ for $x = 0$. The graphs for the arcsine and arccosine can be seen in figure 15.

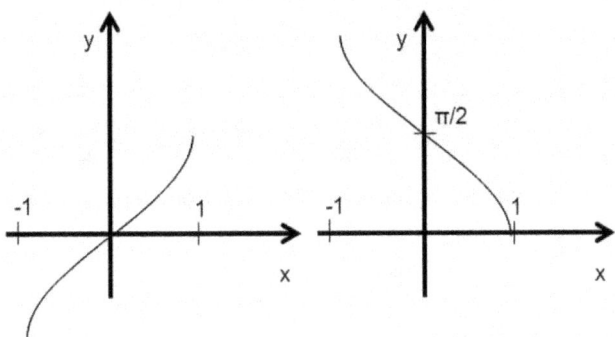

Figure 15: The functions arcsin x (left) and arccos x (right).

The tangent moved toward infinity for $x = \pm\frac{\pi}{2}$. The inverse function is thus only defined in the output interval $\left[-\frac{\pi}{2}, \frac{\pi}{2}\right]$. The arctangent is shown in figure 16.

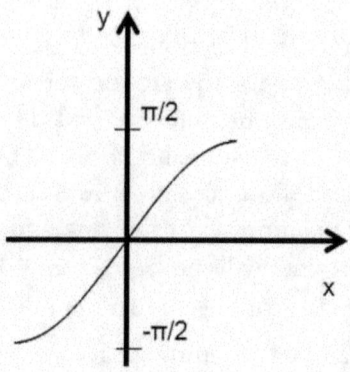

Figure 16: The function arctan x.

After this introduction to the trigonometric functions, we can now take a look at the relationship between the e-function and the trigonometric functions which is the reason for the extensive use of the e-function is physics.

3.4. Euler's Formula

Leonard Euler was working in several mathematical fields. One of the fields was series expansion. A series expansion is a method for calculating a function that cannot be expressed just by elementary operations. An elementary operation is addition, subtraction, multiplication, or division. Functions that cannot be expressed by elementary operations are for example the exponential function, the logarithm or the trigonometric functions.

The series expansion tries to find an expression for such a complicated function that is only based on elementary operations. The goal is to come up with a form like:

$$f(x) = a + bx + cx^2 + dx^3 + ex^4 + \cdots$$

Quite often, such a series can be infinite, i.e. you can only approximate the actual function with certain accuracy (just as we are approximating the number e with the limit of $\left(1 + \frac{1}{n}\right)^n$). And sometimes the final series is not just made up of elementary operations...

One series expansion that is quite often used in physics was first described by the French mathematician Joseph Fourier at the beginning of the 19th century. The Fourier series was originally derived for oscillations with the period T.

The rule that Fourier proved states that every periodic function can be written as a combination of sine and cosine functions. They have the form:

$$\begin{aligned} f(x) = a_0 &+ (a_1 \cos(\omega x) \\ &+ b_1 \sin(\omega x)) \\ &+ (a_2 \cos(2\omega x) + b_2 \sin(2\omega x)) + \cdots \end{aligned}$$

where $\omega = \frac{2\pi}{T}$ is the fundamental frequency of the oscillation with the period T.

A rectangular pulse that has a maximum value of h for half a period and a minimum value of $-h$ for the other half can be written as:

$$f(x) = \frac{4h}{\pi}\left(\sin(\omega x) + \frac{1}{3}\sin(3\omega x) + \frac{1}{5}\sin(5\omega x) + \cdots\right)$$

Figure 17 shows the first three summands of this Fourier series. Just by picking the right sine functions, we can approximate a rectangular pulse fairly well.

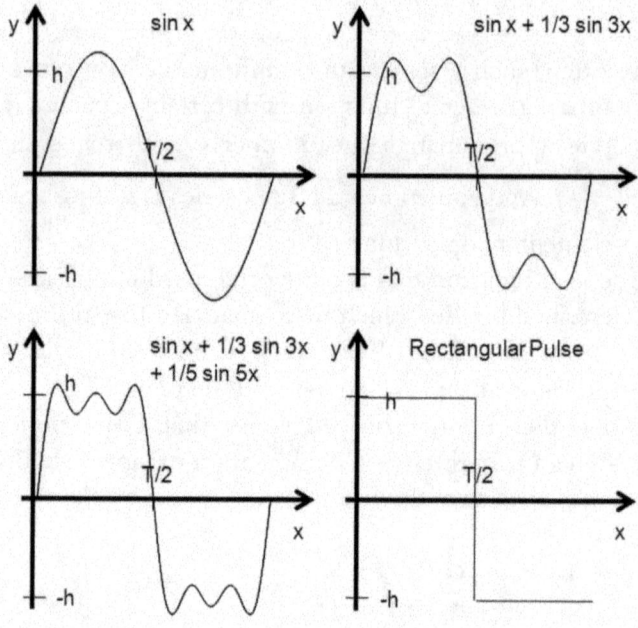

Figure 17: Successive approximation of a rectangular pulse using the Fourier series.

The Fourier series is used quite often in physics where physicists are dealing with periodic behaviors of system whose periodicity, however, cannot always be described easily.

Another important series expansion is the Taylor series which was developed at the beginning of the 18th century by the British mathematician Brook Taylor. This series develops a function f around a point a. The summands of the series are the function and its derivatives. We still consider a derivative to be the slope of a function, we will learn more about it later. The first derivative of a function f

is written as f' the second as f'' and so on. Thus, the Taylor series can be written in a general form as:

$$f(x) = f(a) + \frac{f'(a)}{1!}(x-a) + \frac{f''(a)}{2!}(x-a)^2 + \frac{f'''(a)}{3!}(x-a)^3 + \cdots$$

The formula **3!** stands for the factorial and means $3! = 1 \cdot 2 \cdot 3$. The nth factorial is $n! = 1 \cdot 2 \cdot 3 \cdot \ldots \cdot n$.
The Taylor series for the e-function around the point $a = 0$ is:

$$e^x = 1 + x + \frac{x^2}{2!} + \frac{x^3}{3!} + \cdots$$

This Taylor series is very simple as the derivative of e^x is e^x again, and we have: $e^0 = 1$. This means that we only have to calculate the expression $\frac{(x-a)^n}{n!}$ in the end – for $a = 0$.

Figure 18 shows the Taylor series of the e-function for the first four summands around the point $x = 0$. As you can see, the sum is approximating the e-function better and better.

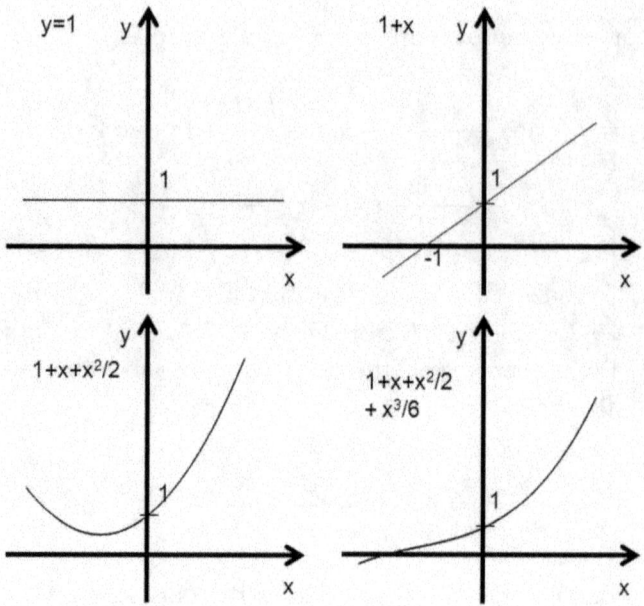

Figure 18: Successive approximation of the e-function using the Taylor-series.

Similarly, we can approximate the sine and cosine around the point of origin by a Taylor series. This results in (more about the derivatives of the trigonometric functions later):

$$\sin x = x - \frac{x^3}{3!} + \frac{x^5}{5!} - \ldots$$
$$\cos x = 1 - \frac{x^2}{2!} + \frac{x^4}{4!} - \ldots$$

This reminds us of the Taylor series of the e-function:

$$e^x = 1 + x + \frac{x^2}{2!} + \frac{x^3}{3!} + \ldots$$

The only problem is that all summands for the Taylor series of the e-function are positive while some summands for the Taylor series of the sine and the cosine are negative. Euler found an e-function that reflected this behavior. Instead of developing e^x into a Taylor series, he used e^{ix}. This gives:

$$e^{ix} = 1 + ix + \frac{(ix)^2}{2!} + \frac{(ix)^3}{3!} + \frac{(ix)^4}{4!} + \frac{(ix)^5}{5!} \ldots$$

The product $i \cdot i$ is simply -1 (the square of $\sqrt{-1}$). Therefore we have:

$$e^{ix} = 1 + ix - \frac{x^2}{2!} - i\frac{x^3}{3!} + \frac{x^4}{4!} + i\frac{x^5}{5!} - \ldots$$

$$e^{ix} = 1 - \frac{x^2}{2!} + \frac{x^4}{4!} - \ldots + i\left(x - \frac{x^3}{3!} + \frac{x^5}{5!} - \ldots\right)$$

$$e^{ix} = \cos x + i \sin x$$

This is Euler's Formula which relates the trigonometric functions sine and cosine with the e-function.

For $x = \pi$ Euler's Formula yields Euler's Identity:

$$e^{i\pi} = -1$$

If you can write the e-function as a combination of trigonometric functions, then you can also write the trigonometric functions as a combination of e-functions.

To write the sine as a combination of e-functions, we start with the difference between two e-functions:

$$e^{ix} - e^{-ix}$$
$$= (\cos x + i \sin x) - (\cos x - i \sin x)$$
$$= 2i \sin x$$

This gives:

$$\sin x = \frac{e^{ix} - e^{-ix}}{2i}$$

The sum of two e-functions gives a formula for the cosine:

$$e^{ix} + e^{-ix}$$
$$= (\cos x + i \sin x) + (\cos x - i \sin x)$$
$$= 2 \cos x$$

Thus:

$$\cos x = \frac{e^{ix} + e^{-ix}}{2}$$

These formulas might look fairly complicated (after all we are combining complex numbers with trigonometric functions and the e-function). But they help to simplify calculations in physics and engineering as the e-function is a very handy function in analysis.

4. Analysis

4.1. The Limit

The Greek philosopher Zeno of Elea who lived in the 5th century B.C. described a paradox of movement. He considered a racing duel between the runner Achilles and a tortoise. The tortoise gets a certain head start, because it is obviously slower than Achilles. At a given sign, they both start to run. After some time, Achilles reaches the point where the tortoise started. But the tortoise has advanced some meters in the meantime. Let's assume that its speed is half of the speed of Achilles. So when Achilles reaches its starting point, it still is half the distance away. Achilles covers this distance again, but the tortoise is still ahead by one fourth of the original distance. When Achilles has covered this distance, it is still one eight ahead. As long as they are both running, the tortoise will keep its lead. There is no way that Achilles can reach the tortoise or even overtake it.

Of course, we know that he does, that's why this riddle is a paradox. But where is our error in reasoning?

Let's write down the distances Achilles is covering while running. The first distance may have the length of one. Then the sum is:

$$1 + \frac{1}{2} + \frac{1}{4} + \frac{1}{8} + \frac{1}{16} + \frac{1}{32} + \ ...$$

We keep on adding a little something to the sum. The distance seems to be infinite, and Achilles can never reach the turtle.

But let's change our point of view and see how far away this sum is from the number 2. When Achilles has covered the first distance, then he still needs to run the same distance to reach the distance of two. When he has covered the first two distances, he still needs to cover half the original distance. After three distances, he still needs to cover one fourth of the original distance, then one eight, one sixteenth and so on. The difference to the distance of two gets smaller and smaller – but it will never be zero.

We can express this in a different way: The series above approaches the number 2. After the first summand, we still need a 1 to reach the 2. Then we need $\frac{1}{2}$, then $\frac{1}{4}$, then $\frac{1}{8}$, then $\frac{1}{16}$ and so on. The difference to 2 will become smaller with each summand, but we will never reach the 2.

Even though this series has an infinite number of summands, it does not become infinitely large but is limited by a finite number, the number 2.

And this is the error of reasoning that Zeno made. Even though he summed up an infinite number of distances, the overall distance was not infinite, but finite – and could thus be covered by Achilles in a finite time. So he has to overtake the tortoise, as, in fact, he does.

Mathematicians say that the series

$$1 + \frac{1}{2} + \frac{1}{4} + \frac{1}{8} + \frac{1}{16} + \frac{1}{32} + \ ...$$

reaches the limit of 2.

We can write the series in a more compact way using the sigma sign Σ:

$$1 + \frac{1}{2} + \frac{1}{4} + \frac{1}{8} + \frac{1}{16} + \frac{1}{32} + \ldots = 1 + \sum_{n=1}^{\infty} \frac{1}{2n}$$

The notation at the sigma sign means that the summation starts with $n = 1$ and goes up to infinity.

The limit of the sum (designated with the symbol "lim" for the Latin word "limes") is:

$$\lim_{n \to \infty} \left(1 + \sum_{n=1}^{\infty} \frac{1}{2n} \right) = 2$$

Now wait. We just said that this series has a limit of 2, but never reaches it. Nevertheless we use the equal sign? This is correct, because we can make the error as small as we want. Mathematicians say that we can find an arbitrary number ε which is larger than the difference between the limit of the sum and the sum itself:

$$\varepsilon > 2 - \left(1 + \sum_{n=1}^{\infty} \frac{1}{2n} \right)$$

Some summands may lay outside of the error margin ε, but the majority lies well within (as there is an infinite number of summands). If we have $\varepsilon = \frac{1}{8}$, then the summands $1, 1\frac{1}{2}, 1\frac{3}{4}$ and $1\frac{7}{8}$ are outside of this limit. The number $1\frac{15}{16}$ would be the first to meet the error margin – and then there is an infinite number of other numbers. Thanks to this definition of the limit (the error between the limit and the sum can be made arbitrarily small) we can use the equal sign even though the limit is never reached.

We can define a limit for a function in a similar way. However, in the case of a function we have to consider that the value of the function $f(x)$ depends on the variable x. That means that we have to assure that the difference of the function at $f(x)$ around $f(x_0)$ is smaller than ε, and at the same time, there has to be a region δ around x_0, where all x-values result in a $f(x)$ which lies in the ε-region. This is shown in figure 19.

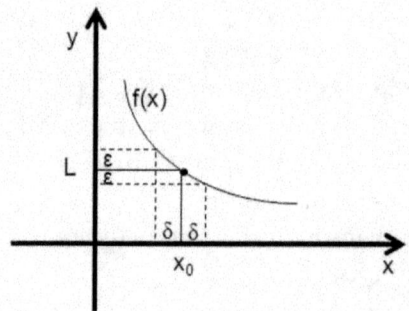

Figure 19: The limit of a function: To each ε-region there is a respective δ-region so that all function values of that region are in the ε-region.

The definition of the limit requires that we can define a small ε-region where we can find the values of the function $f(x)$ as well as the limit L. But we also need a δ-region around the value x_0 which belongs to the limit L so that all x-values from this δ-region will result in function values $f(x)$ in the ε-region. If there are x-values of the δ-region that result in $f(x)$-values outside of the ε-region, then we are not able to define a limit. Figure 20 shows such a function. This function "jumps" around the x value x_0, and we cannot define a limit for $f(x)$ at x_0.

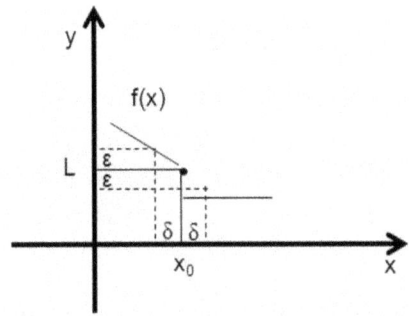

Figure 20: In the case of a discontinuous function, we cannot find to each ε a δ-region so that all function values from that region are within the ε-region. A limit cannot be defined at this point.

As the function shown in figure 20 jumps around the value x_0, we cannot find an arbitrary ε-region around the limit L so that all x-values from the δ-region around x_0 result in function values $f(x)$ that lie in the ε-region. Therefore, this function does not have a limit at the point (x_0, L).

Written in a mathematically correct way, we can define the limit of a function as follows:

A function $f(x)$ has the limit L for $x \to x_0$ if we have a $\delta > 0$ for each $\varepsilon > 0$, so that for all x-values that fulfill the condition $0 < |x - x_0| < \delta$ also $|f(x) - L| < \varepsilon$ is valid.

The vertical brackets $|x|$ mean that we only use the absolute values of x, i.e. we ignore all negative signs as all negative numbers would be smaller than any small positive number per definition.

With this definition of the limit, we are now prepared to move on to differential calculus.

4.2. Differential Calculus

Quite often we would like to know how a function behaves. The function might be speed for example, and we would like to know how the speed of a car changed over time, i.e. what the acceleration was at different times.

If you have a graph of the speed of your car, the acceleration, the change of speed, is nothing but the slope of the curve at the interesting point (if the speed doesn't change, the slope is zero; if the speed increases than the slope increases – and this increase is steeper if the acceleration is greater).

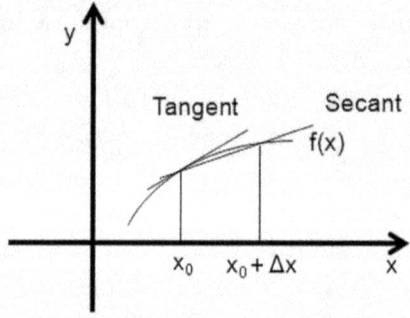

Figure 21: The slope of a curve f(x) at the point x_0, the tangent, can be approximated by a secant through the points x_0 and $x_0 + \Delta x$.

The slope of a graph at x_0 is nothing else but the tangent of the graph at x_0. If you want to know the slope of a graph at x_0 you only have to calculate the tangent at this point.

That is easier said than done. For a starting point, we can try to calculate the secant at the point x_0 and a neighboring point $x_0 + \Delta x$, as it is shown in figure 21. The slope s of the secant is the change of height divided by the length. The change of height is $f(x_0 + \Delta x) - f(x_0)$, the length is $x_0 + \Delta x - x_0$, i.e. we have:

$$s = \frac{f(x_0 + \Delta x) - f(x_0)}{x_0 + \Delta x - x_0}$$

$$= \frac{f(x_0 + \Delta x) - f(x_0)}{\Delta x}$$

The slope of the secant might be close to the slope of the tangent, but it is not quite the slope of the tangent. You would only get if, if Δx approaches zero (if Δx is zero we have an insoluble equation).

Now is the time to use the limit we discussed in the previous chapter. The tangent can be defined as the limit of the secant at x_0 for Δx approaching zero.

If such a limit exists, we say that the function $f(x)$ can be differentiated. We can thus calculate the derivative of a function if

$$\lim_{\Delta x \to 0} \frac{f(x_0 + \Delta x) - f(x_0)}{\Delta x}$$

exists.

There are several ways to write down the derivative of $f(x)$ for x at the point x_0:

$$f'(x_0) \text{ or } \left.\frac{df(x)}{dx}\right|_{x=x_0} \text{ or } \frac{df}{dx}(x_0) \text{ or } \frac{d}{dx}f(x_0)$$

We can have the case that a function depends on several variables but we are only interested in differentiating it with respect to one variable. In this case we write:

$$\frac{\partial f(x_1, \ldots, x_n)}{\partial x_i}$$

This means that the function f, which depends on the variables x_1 to x_n, will be differentiated with respect to the variable x_i. Such derivatives are called partial derivatives as the function will only be partially differentiated.

The derivatives of derivatives are designated with higher exponents, i.e. we write the second derivative as:

$$f''(x_0) \text{ oder } \frac{d^2f}{dx^2}$$

Let us now calculate the derivative for a simple function. The function shall be the function of a parable: $f(x) = 2x^2 + 3$.

We start with calculating the slope of the secant at the points x_0 and $x_0 + \Delta x$:

$$\frac{f(x_0 + \Delta x) - f(x_0)}{\Delta x}$$

$$= \frac{2(x_0 + \Delta x)^2 + 3 - (2x_0^2 + 3)}{\Delta x}$$

$$= \frac{2(x_0^2 + 2x_0\Delta x + \Delta x^2) + 3 - 2x_0^2 - 3}{\Delta x}$$

$$= \frac{2x_0^2 + 4x_0\Delta x + 2\Delta x^2 - 2x_0^2}{\Delta x}$$

$$= 4x_0 + 2\Delta x$$

We get the derivative if we calculate the limit for $\Delta x \to 0$:

$$\frac{d}{dx}f(x_0)$$

$$= \lim_{\Delta x \to 0}(4x_0 + 2\Delta x)$$

$$= 4x_0$$

So the derivative for the function $f(x)$ in general is: $f'(x) = 4x$.

For many functions we can find the derivatives in large tables. For power functions, we have the simple rule:

$$\frac{d}{dx}x^n = n \cdot x^{n-1}$$

We can prove this with the definition of the limit:

$$\frac{(x + \Delta x)^n - x^n}{\Delta x}$$

$$= \frac{x^n + nx^{n-1}\Delta x + (n-1)x^{n-2}\Delta x^2 + (n-2)x^{n-3}\Delta x^3 + \cdots - x^n}{\Delta x}$$

$$= nx^{n-1} + (n-1)x^{n-2}\Delta x + (n-2)x^{n-3}\Delta x^2 + \cdots$$

For $\Delta x \to 0$ all summand but the first disappear and we have:

$$\frac{d}{dx}x^n = n \cdot x^{n-1}$$

The derivation of the e-function can be calculated with the same approach:

$$\frac{e^{x+\Delta x} - e^x}{\Delta x}$$

$$= \frac{e^x \cdot e^{\Delta x} - e^x}{\Delta x}$$

$$= e^x \cdot \frac{e^{\Delta x} - 1}{\Delta x}$$

We want Δx to approach zero, so we have to figure out what the limit for this fraction is. The numerator and the denominator are both approaching zero. We could argue that the e-function behaves linearly at $x = 0$, i.e. the numerator and the denominator are approaching zero with the same "speed", so both are more or less equal and we get the limit 1. As we will see, the result is correct. The argumentation, however, is shaky. To get a solid proof, we set $e^{\Delta x} = 1 + u$. Then we have $\Delta x = \ln(1+u)$ and we get:

$$\lim_{\Delta x \to 0} \frac{e^{\Delta x} - 1}{\Delta x}$$

$$= \lim_{u \to 0} \frac{u}{\ln(1+u)}$$

$$= \lim_{u \to 0} \frac{1}{\frac{\ln(1+u)}{u}}$$

For the natural logarithm, we have $\ln x \leq x - 1$ (at $x = 1$ the straight line touches the natural logarithm from above). And we have $\ln x \geq 1 - \frac{1}{x}$ (the function touches the natural logarithm from below at $x = 1$, see figure 22).

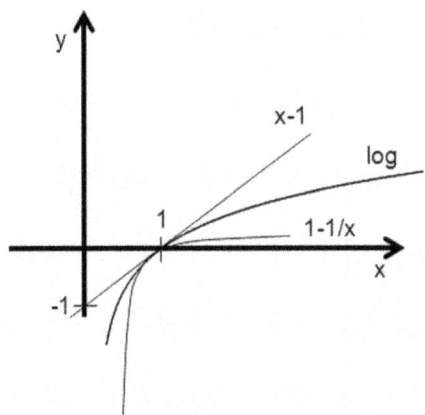

Figure 22: The logarithm is surrounded by the functions f(x)=x-1 and f(x)=1-1/x.

Thus we have the inequality:

$$1 - \frac{1}{x} \leq \ln x \leq x - 1$$

If we set $x = 1 + u$, then we get:

$$1 - \frac{1}{1+u} = \frac{u}{1+u} \leq \ln(1+u) \leq 1 + u - 1$$

If we divide the inequality by u, we get:

$$\frac{1}{1+u} \leq \frac{\ln(1-u)}{u} \leq 1$$

The middle part of this inequality is the denominator of the expression

$$\lim_{u \to 0} \frac{1}{\frac{\ln(1+u)}{u}}$$

the limit of which we are trying to calculate.

If u approaches zero in the inequality, then the lower limit approaches one. The upper limit is one anyway, i.e. the expression $\frac{\ln(1+u)}{u}$ has to approach 1 as well.

This means, we also have

$$\lim_{\Delta x \to 0} \frac{e^{\Delta x} - 1}{\Delta x} = 1$$

With this, we have shown that the derivative of e^x is e^x again:

$$\frac{d}{dx} e^x = e^x$$

Some other often used derivatives are (now without a proof):

$$\frac{d}{dx} \ln x = \frac{1}{x}$$

$$\frac{d}{dx} \sin x = \cos x$$

$$\frac{d}{dx} \cos x = -\sin x$$

The fact that the sine and cosine are each other's derivative (with a minus sign appearing from time to time) explains the Taylor series for these two function that we wrote down when deriving Euler's formula. The Taylor series only showed every second summand (either only the even or the odd exponents). As we where developing the series around the point of origin, the summands disappeared for the sine and were one for the cosine.

You can find derivatives for a lot of function in tables. But sometimes we have to deal with combinations of functions g and h. In this case, we can use the following rules:

1) Sum rule: $(g \pm h)' = g' \pm h'$
2) Product rule: $(g \cdot h)' = g' \cdot h + g \cdot h'$
3) Quotient rule: $\left(\frac{g}{h}\right)' = \frac{g' \cdot h - g \cdot h'}{h^2}$
4) Chain rule: $(g(h(x))' = g'(h(x)) \cdot h'(x)$

We will prove the product rule as an example. The trick is, again, to expand the expression by adding a zero that has been chosen carefully. We start with:

$$\frac{d}{dx}(g \cdot h)$$

$$= \lim_{\Delta x \to 0} \frac{g(x+\Delta x) \cdot h(x+\Delta x) - g(x) \cdot h(x)}{\Delta x}$$

Now we add zero in the form of:

$$0 = \frac{g(x) \cdot h(x+\Delta x)}{\Delta x} - \frac{g(x) \cdot h(x+\Delta x)}{\Delta x}$$

and get:

$$\frac{d}{dx}(g \cdot h)$$

$$= \lim_{\Delta x \to 0} \frac{g(x+\Delta x) - g(x)}{\Delta x} \cdot h(x+\Delta x)$$
$$+ \lim_{\Delta x \to 0} g(x) \cdot \frac{h(x+\Delta x) - h(x)}{\Delta x}$$

If Δx approaches zero, the fractions are nothing but the respective derivatives and $h(x+\Delta x)$ will be $h(x)$. Thus, we have the product rule:

$$\frac{d}{dx}(g \cdot h) = \frac{d}{dx}g \cdot h + g \cdot \frac{d}{dx}h$$

or

$$(g \cdot h)' = g' \cdot h + g \cdot h'$$

We can use the derivatives to learn more about a graph without having to draw it (which wouldn't be possible anyway as most functions are defined from $-\infty$ to $+\infty$). Derivatives can help to find the extreme values (minima and maxima) and the point where a graph changes its slope (the so-called inflection point). Especially finding extreme points can help to optimize a process, for example if you want to figure out the area that has a maximum size for a given circumference.

How can we see if a function has a minimum or a maximum? Let's take the function

$$f(x) = \frac{1}{3}x^3 - 2x^2 + 3x$$

as an example. The function is shown in figure 23.

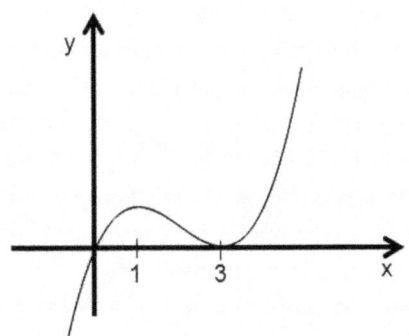

Figure 23: The function $f(x) = 1/3\ x^3 - 2x^2 + 3x$

We can see at once that the function is lying above the x-axis for positive x-values, and that the function has smaller and bigger values for positive x-values. The function has a local maximum for $x = 1$ and a local minimum for $x = 3$. We can find these extreme points very easily just by looking at the function. However, it will be more complicated for other functions. We have to find a rule that tells us where we can find extreme points. Looking at the figure gives us an idea for the approach: If we have a maximum, the graph rises and declines right after it. The slope is going from positive to negative around a maximum and is zero at the maximum. Just before a minimum, the slope is negative, at the minimum it is zero and right afterwards it is positive. The derivative describes the slope of a function, i.e. the necessary condition for an extreme point is:

$$f'(x) = 0$$

The derivative of our example is:

$$f'(x) = x^2 - 4x + 3$$

This equation is solved for x = 1 and x = 3, the respective values of the function are $\frac{4}{3}$ and 0. The graph has horizontal tangents at the points $(1, \frac{4}{3})$ and $(3, 0)$.

The question now is if we really have an extreme point there – and if this is the case if this point is a maximum or a minimum.

We have to answer the question if we really have an extreme point at these positions as a horizontal tangent doesn't necessarily mean that there really is an extreme point. Let's take a look at the function $f(x) = x^3$ which is shown in figure 24. This function obviously has a horizontal tangent at $x = 0$ (as the derivative $f'(x) = 3x^2$ is zero at this point), but the function obviously does not have a maximum or minimum at this point. Such a point is called a saddle point.

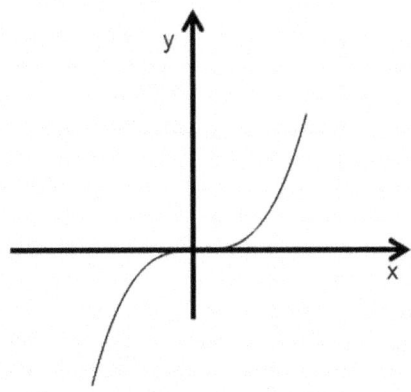

Figure 24: The function f(x) = x³. This function has a horizontal tangent at the point (0, 0), but no extreme point. This point is called saddle point.

We need an additional criterion that helps us to recognize an extreme point. To find this, we take a closer look at the derivative of $f(x) = \frac{1}{3} x^3 - 2x^2 + 3x$. The derivative is shown in figure 25.

We can see that the derivative (a parable) has no extreme points at the extreme points of $f(x)$, i.e. the second derivative of $f(x)$ is not zero at these points. On the other hand, the derivative of $f(x) = x^3$, the function $f'(x) = 3x^2$ has a horizontal tangent at the point (0, 0) again, after all the second derivative $f''(x) = 6x$ is zero for x = 0.

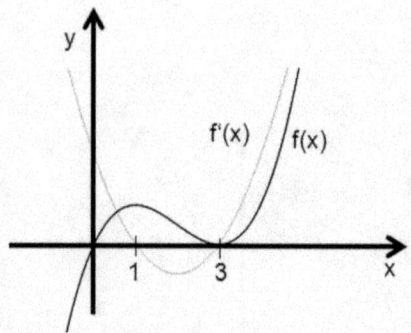

Figure 25: The function f(x) = 1/3 x³ − 2x² + 3x and its derivative f'(x) = x² − 4x + 3.

In addition to the necessary condition that we need $f'(x) = 0$, we also need to fulfill the sufficient condition that we have $f''(x) \neq 0$ to get a local extreme point. If we have $f''(x) = 0$, then we have a local saddle point.

Furthermore, we can see from figure 25 that we have a maximum if the slope of the derivative is negative, i.e. is we have $f''(x) < 0$. If we look at the original function $f(x)$, then we see that it has a positive slope before a maximum and a negative slope afterwards. The derivative "falls" from positive to negative values, its slope is negative.

We also see that the slope of the derivative at a minimum has to be positive, i.e. we have a minimum if the condition $f''(x) > 0$ is fulfilled.

In addition to the extreme points, the minima and maxima, we often also like to know where the function changes its slope, i.e. where we have inflection points.

As the function $f(x) = \frac{1}{3} x^3 - 2x^2 + 3x$ has a minimum and a maximum it has to change its slope somewhere. The inflection point has to be somewhere between the maximum and the minimum; because for x-values larger than zero the function bends to the right, in clockwise direction, climbs up to a maximum and falls down again. But instead of going to smaller and smaller values, the function changes its slope. The graph bends to the left, anti-clockwise, goes through a minimum and rises again. The inflection point can thus be found at the point where the graph changes its slope from falling to rising.

To find the inflection point, we thus have to look at the slope of the slope, i.e. the slope of the derivative. We have an inflection point where this slope changes its direction. The derivative has an extreme point here, i.e. the slope of the derivative is zero. The slope of the derivative is given by the second derivative. We thus have an inflection point when the second derivative is zero:

$$f''(x) = 0$$

This means for our example that

$$f''(x) = 2x - 4$$

has to be zero. This happens for $x = 2$. Figure 25 shows that the derivative has a minimum at this point.

But just as in the case of the extreme points, the condition $f''(x) = 0$ is only a necessary condition that we have an inflection point, i.e. this condition has to be fulfilled or we cannot have an inflection point at all. But this condition is not sufficient in the sense that every point that fulfills this condition will be an inflection point.

In the case of the extreme points we had seen that the necessary condition could also give us a saddle point instead of an extreme point. Only with the sufficient condition $f''(x) \neq 0$ we could be sure that the calculated point was really an extreme point.

We have the same problem for the inflection point: The derivative has to have an extreme point; otherwise we do not have an inflection point. So the second derivative must not equal zero – but this time it is the second derivative of the derivative, i.e. we have the sufficient condition:

$$f'''(x) \neq 0$$

With these necessary and sufficient conditions we can calculate the main characteristic points of a function.

4.3. Integral Calculus

If a graph is given, you might want to know what the slope of the graph is or where the graph has extreme points. But you might also want to know the size of the area below that graph. The area of a triangle or a circle can be calculated quite easily with compact formulas. But what if we have an arbitrary circumference?

For simplicity, we can ask what the size of the area between a graph given by the function $f(x)$ and the x-axis is. If the lower side of the body would be arbitrary as well, then we could calculate the area between the function $g(x)$ that describes the lower graph and the x-axis and subtract it from the area below the function $f(x)$.

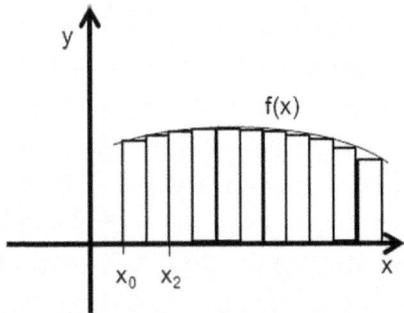

Figure 26: We can approximate the area below the function f(x) with inscribed rectangles.

We can try to approximate the area of an arbitrary function with rectangles. Figure 26 shows rectangles that are all located below the graph of the function $f(x)$. The area of the kth rectangle is given by:

$$(x_k - x_{k-1}) \cdot \text{Min}_k (f(x))$$

Here $\text{Min}_k (f(x))$ is the smallest value for $f(x)$ in the kth rectangle. We get an approximation for the area if we sum up all the rectangles:

$$\sum_{k=1}^{n} (x_k - x_{k-1}) \cdot \text{Min}_k (f(x))$$

We could also take the maximum $\text{Max}_k (f(x))$ of the function $f(x)$ in the kth rectangle, or any value of $f(x)$ in the rectangle like the value at the corner-point x_k. If we set $\Delta x_k = (x_k - x_{k-1})$, then the sum is:

$$\sum_{k=1}^{n} f(x_k) \cdot \Delta x_k$$

The error that we make with this approach is getting smaller the smaller the rectangles are. And again we need the omnipresent limit. We get the area below the function $f(x)$ if Δx approaches zero and we sum up all rectangles below the graph. The result of this summation is written as an integral:

$$\int_{x_0}^{x_n} f(x)dx$$

This symbol means that we integrate the function $f(x)$ between the values x_0 and x_n. Even though our descriptive derivation of the integral would mean that we are summing up the infinitesimally small pieces with the width dx (the integral sign is nothing but an abstract sigma sign), the integral shall not be understood as a summation and there is no multiplication taking place between $f(x)$ and dx. The dx is only seen as a part of the integral sign and it shows us according to which variable we have to perform the integration of the function. For it could be that the function depends on several variables, but we only want to integrate it over one variable (e.g. x). So we would write:

$$\int_{a}^{b} f(x,y)dx$$

It can also be important in which order we are performing the integration if the function depends on several variables. We always start with the inner variable, and both notations

$$\int_a^b \int_c^c f(x,y)dxdy = \int_a^b dy \int_c^d dx\, f(x,y)$$

can be used equally. In both case we first integrate over x from c to d.

But what does it mean if we want to integrate a function? How can we do this without summing up the rectangles and make the width of the rectangles approach zero?

The answer is given by the fundamental theorem of calculus. To derive this, we define the primitive $F(x)$:

$$F(x) = \int_c^x f(t)dt$$

I.e. we get the primitive if we integrate the function f from c to the variable limit x.

Now we derive this primitive:

$$F'(x_0) = \lim_{\Delta x \to 0} \frac{F(x_0 + \Delta x) - F(x_0)}{\Delta x}$$

The difference of the primitive is:

$$F(x_0 + \Delta x) - F(x_0) = \int_c^{x_0+\Delta x} f(t)dt - \int_c^{x_0} f(t)dt$$

Up to now we have only used the definition of the primitive. We also have the following definition:

$$\int_a^b f(t)dt = -\int_b^a f(t)dt$$

This means that the area below a function shall be negative if we exchange the limits (i.e. we sum up the rectangles from the other side).

In addition, we have:

$$\int_a^b f(t)dt + \int_b^c f(t)dt = \int_a^c f(t)dt$$

If we calculate the area between the limits $[a, b]$ and add the area between $[b, c]$ to this, we could have calculated the area between $[a, c]$, i.e. the whole range, instead.

With these two formulas in mind we take a look at the difference of the primitives:

$$F(x_0 + \Delta x) - F(x_0) = \int_c^{x_0+\Delta x} f(t)dt - \int_c^{x_0} f(t)dt$$

$$= \int_c^{x_0+\Delta x} f(t)dt + \int_{x_0}^c f(t)dt$$

$$= \int_{x_0}^{x_0+\Delta x} f(t)dt$$

That means we can write the derivative of the primitive as

$$F'(x_0) = \lim_{\Delta x \to 0} \frac{1}{\Delta x} \int_{x_0}^{x_0+\Delta x} f(t)dt$$

This still looks rather complicated. But the integral only means that we want to calculate the area below the function $f(t)$ for the region Δx. If this is possible (which we assume) then there is a rectangle with the height $f(\xi)$ (ξ is the Greek letter xi) and the width Δx which is exactly as big as the area below the graph (the value ξ is taken from the interval $\Delta x = [x_0, x_0 + \Delta x]$). This has to be the case as we can find a rectangle $f(x_i)\Delta x$ which is bigger and another rectangle which is smaller. As we are working with limits, the function has to be smooth and cannot make jumps; i.e. there has to be a rectangle between the bigger and the smaller one that is exactly as big as the area below the function, i.e. we have:

$$\int_{x_0}^{x_0+\Delta x} f(t)dt = f(\xi) \cdot \Delta x$$

What the variable ξ looks like and thus the function-value f(ξ) is of no interest at all, because our intention is anyway that Δx shall approach zero. Then ξ is approaching x_0, after all the interval has no width left whatsoever.

So we have

$$F'(x_0) = \lim_{\Delta x \to 0} \frac{F(x_0 + \Delta x) - F(x_0)}{\Delta x}$$

$$= \lim_{\Delta x \to 0} \frac{1}{\Delta x} f(\xi) \cdot \Delta x = \lim_{\Delta x \to 0} f(\xi)$$

With Δx approaching zero, we get the remarkable result:

$$F'(x) = f(x)$$

The is the fundamental theorem of calculus: The derivative of the integral of a function $f(x)$ is the original function $f(x)$. Differentiation is the inverse of integration.

As we have large tables that give us the derivatives of functions, we only have to take a look at the derivatives to find the primitive of a given function.

But there is one imprecision that we have made when defining the primitive, as we can see now: The derivative of the primitive is the function that we want to integrate. This means we need to add a constant summand to the primitive. For if we derive the primitive the constant summand will vanish. The exact definition of the primitive thus has to be:

$$\int_a^x f(t)dt = F(x) + C$$

We derived the fundamental theorem of calculus by using the definition of the derivative without taking into consideration this constant summand. But as we were calculating the difference between two primitives, the constant summand C would have disappeared anyway, so the theorem is still correct.

The value of C is not arbitrary but can be calculated easily. To get this value, we simply set $x = a$ and calculate the integral for the simple case of an interval going from a to a. As the area under the function has to be zero (after all the width is zero), we get:

$$\int_a^a f(t)dt = 0 = F(a) + C$$

and thus

$$C = -F(a)$$

The primitive is therefore:

$$\int_a^x f(t)dt = F(x) - F(a)$$

An integral from a to b for the function $f(x)$ is then:

$$\int_a^b f(t)dt = F(b) - F(a)$$

Let's calculate a concrete example. Our function is $\frac{1}{x}$, and we want to integrate it from $a = 1$ to $b = 10$:

$$\int_1^{10} \frac{1}{x}\, dx$$

Looking into tables (or just some pages back), we find that $\frac{1}{x}$ is the derivative of $\ln x$. Thus we have

$$\int_1^{10} \frac{1}{x}\, dx = \ln x \big|_1^{10}$$

As an intermediate, some mathematicians write the primitive in its general form and note the limits at a vertical line. The calculation finally yields:

$$\int_1^{10} \frac{1}{x}\, dx = \ln 10 - \ln 1 = 2.3\ldots - 0 \approx 2.3$$

However, by far not all integrals can be calculated that easily. This direct calculation is called analytical calculation. In most cases we have so solve the integral numerically with the help of a computer – where we sum up very small rectangles under the function again.

4.4. Differential Equations

We sometimes have to deal with an equation which not only contains a function but also one of its derivatives (or even several of them). One simple example for this is the spring pendulum that is shown in figure 27.

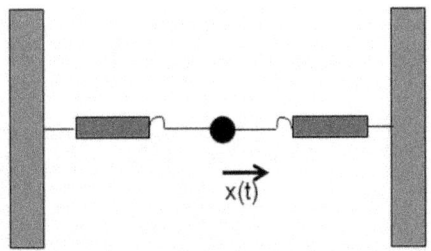

Figure 27: The spring pendulum. The relation between extension x and force F is given by Hooke's law: F = - Dx.

The force F which elongates the spring pendulum is proportional to the extension x of the pendulum. The proportionality is described by Hooke's law, using the constant $-D$.

$$F = -Dx$$

Newton tells us that the force is proportional to the acceleration a with the mass m of the body as the proportionality constant:

$$F = ma$$

The acceleration is nothing but the change of the speed v with time. For small time periods, this is simply the derivative of the speed with respect to the time:

$$F = m\frac{dv(t)}{dt}$$

Speed is nothing but the change of a distance x with time. This means that the acceleration is the second derivative of the distance x with respect to the time t:

$$F = m \frac{d^2 x(t)}{dt^2}$$

This gives us the following equation for the spring pendulum:

$$m \frac{d^2 x(t)}{dt^2} = -D x(t)$$

If we want to know the movement of the spring pendulum with time we have to solve an equation that contains the original function $x(t)$ and its second derivative $\frac{d^2 x(t)}{dt^2}$. We have to find a function $x(t)$ which solves this equation. Equations of this kind which contain a function and its derivatives are called differential equations. As the derivative only depends on one variable, this equation is called an ordinary differential equation. If we set $\omega^2 = \frac{D}{m}$, this differential equation has the form:

$$\frac{d^2 x(t)}{dt^2} + \omega^2 x(t) = 0$$

You might not believe it, but this equation has a very simple solution. Let's just remember the derivatives we have already talked about. The sum of a function with is second derivative has to be zero. This means that the second derivative has to be the negative of the original function. Functions that fulfill this condition are the sine and the cosine. The unknown solution must therefore have the form:

$$x(t) = a \sin \omega t + b \cos \omega t$$

The function $\sin \omega t$ is differentiated using the chain rule, i.e. we calculate the derivative of the function ($\sin x$) and multiply it with the derivative of the function (ωt). This gives: $\frac{d}{dt} \sin \omega t = \omega \cos \omega t$.

The second differentiation adds a second ω (now we understand why we have set $\frac{D}{m} = \omega^2$), and the differential equation is solved.

At least the equation is solved in principle. We still have to figure out what the constants a and b look like. We cannot answer that question yet. We can only answer it if we get additional information. We can have the initial conditions that the displacement was $x = 1$ and the speed was $\frac{dx}{dt} = 2\omega$ at the time $t = 0$.

We can put this information into the original function

$$x(t) = a \sin \omega t + b \cos \omega t$$

and its derivative:

$$\frac{dx(t)}{dt} = a\omega \cos \omega t - b\omega \sin \omega t$$

This gives us (as $\sin 0 = 0$):

$$1 = a \cdot 0 + b \cdot 1$$
$$2\omega = a\omega - b \cdot 0$$

The constants for this case are $b = 1$ and $a = 2$ and the solution is:

$$x(t) = 2\sin\omega t + \cos\omega t$$

Instead of the initial conditions at the time $t = 0$ we could have used the conditions of the system at any time to get the coefficients a and b. They will, of course, be different for different conditions.

A general way to solve a homogenous differential equation of the form

$$f^{(n)}(x) + a_{n-1}f^{(n-1)}(x) + \ldots + a_1 f'(x) + a_0 f(x) = 0$$

is to solve it with an exponential function of the form $f(x) = e^{\lambda x}$. The superscript (n) designates the derivative (n^{th} derivative) and not the nth power (therefore we set it in brackets).

If we put this approach in the homogenous differential equation, then we get (this example again shows how easy to handle the e-function is, as it is its own derivative):

$$\lambda^n e^{\lambda x} + a_{n-1}\lambda^{n-1}e^{\lambda x} + \ldots + a_1 \lambda e^{\lambda x} + a_0 e^{\lambda x} = 0$$

We divide this by $e^{\lambda x}$ and get the auxiliary equation of the differential equation:

$$\lambda^n + a_{n-1}\lambda^{n-1} + \ldots + a_1 \lambda + a_0 = 0$$

Instead of looking for a function, we are now looking for the roots of an equation, i.e. a simple number. This will

then tell as which combination of e-functions will solve the differential equation.
In our example

$$\frac{d^2x(t)}{dt^2} + \omega^2 x(t) = 0$$

We get the auxiliary equation:

$$\lambda^2 + \omega^2 = 0$$

This gives us:

$$\lambda^2 = -\omega^2$$
$$\lambda = \pm i\omega$$

The solution is a complex number. This gives us the complex function:

$$x(t) = e^{i\omega t}$$

(or $x(t) = e^{-i\omega t}$ which makes no difference in the end). Using Euler's formula this can be written as

$$x(t) = \cos \omega t + i \sin \omega t$$

We could argue that we can ignore the imaginary part as we want to describe a behavior in the real world. But a quick calculation shows that the sine could also be a valid solution of the differential equation. The approach with the e-function does not give us the correct solution; it only shows us which functions could possibly solve the differential equation. In our case we get the answer that

$\cos \omega t$ and $\sin \omega t$ could be solutions of the differential equation. The linear combination of these two functions

$$x(t) = a \sin \omega t + b \cos \omega t$$

gives us the correct solution that we had only "guessed" earlier.

The situation is much more complicated if we do not have a differential equation with a function that depends on just one variable, but with functions that depend on several variables. As such equations have partial derivatives the equations are called partial differential equations.

Such an equation can have the form:

$$\frac{\partial f(x,t)}{\partial x} + \frac{\partial f(x,t)}{\partial t} = 0$$

Solving such a differential equation analytically is much harder than solving a simple differential equation. In many case, we don't even try but use a computer to calculate it numerically and show us the graph on the screen. One numerical approach to solve such a partial differential equation is the finite element method.

This method cannot be described in all its details in this short introduction. It was only developed in the 1960s and 1970s when computers were available that were powerful enough to deal with the huge amount of data.

The principle approach is to divide the area of calculation in small elements. For a one-dimensional problem this could be the parts of a line, for a two-dimensional problem this could be triangles or rectangles that cover the plane. For problems in higher dimensions, these elements are basic elements of higher dimensions – which can make it hard to imagine.

The wanted function $f(x)$ is approximated in these basic elements with basic functions $\Phi(x)$, which should be rather simple. The approximated function $f_h(x)$ is written as a linear combination of these basic functions

$$f_h(x) = \sum_{i=1}^{N} u_i \Phi_i(x)$$

Quite often, these partial differential equations are needed if we want to calculate mechanical structures, the heat conductance or electrical or magnetic phenomena. The function is thus restricted on a very well limited area, i.e. we have boundary conditions. In addition, there should not be a jump between the function of one basic element and the functions of its neighboring elements. This gives us further boundary conditions between the basic elements.

If we consider all boundary conditions, we end up with N systems of equation with N unknowns – which can only be solved by a computer.

The finite element method makes it possible to simulate stress in mechanical elements (or even whole buildings) or the distribution of electrical fields very accurately. The method is so powerful thanks to modern computer technology that for instance car manufactures simulate the behavior of cars on the computer with this method before they actually build a model and perform an expensive crash-test.

5. Stochastic

The science of stochastic can be traced back to the ancient times when people started to think about this "art of speculation" (which is the literal translation of the Greek word). For even our ancestors liked to play. But what was the probability that you could get a certain combination of numbers when playing dice? What was the probability to win in a lottery?
Another question came from the area of insurance. If merchant ships with valuable goods were travelling through the Mediterranean Sea – what was the probability that they would reach the port of their destination unharmed?
These are the questions that stochastic deals with. It is subdivided into the fields of statistics and the calculus of probabilities.

5.1. Statistics

Statistics is the science of collecting data and evaluating them. Originally, this only meant data that concerned the state. After all, the word statistics is derived from the Latin word "statisticum" which means state-related.
In ancient times the state was probably the only institution that was interested in collecting big amount of data – after all it wanted to know how much taxes it could collect. Today, also private companies are very much interested in private data, for instance to send us personalized ads. Thus, statistics today just means collecting and evaluating data.

If you have a big amount of data, for instance the height of buildings in a city, then this is just a huge amount of information that needs some sorting. You could for instance write down the height of each building, or you could get an overview by just calculating the average height. Let's assume the buildings in a city have the heights of 5m, 6m, 5m, 4m, 4m, 40m, and 6m – we only have some houses and one church. The average (the arithmetic average) is calculated by summing up the heights of all buildings and dividing that number by the number of buildings. The average is calculated using the formula:

$$\bar{x} = \frac{1}{n} \sum_{i=1}^{n} x_i$$

In our case we have:

$$\bar{x} = \frac{1}{7}(5 + 6 + 5 + 4 + 4 + 40 + 6)m = 10m$$

The average of the buildings in the city is 10m. This gives the impression that all buildings in the city are very tall, whereas in fact there is only one very tall building (the church) and all other buildings are significantly smaller.

So we need a measure that tells us how well the collection of data is represented by the average, i.e. we have to calculate the deviation from the average. The deviation is calculated with its absolute value (we just want to know its size), i.e. it is calculated as $|x_i - \bar{x}|$.

However, it is not very easy to perform calculations with absolute values. Therefore, we generally use the square of this difference $(x_i - \bar{x})^2$ as a measure for the deviation. If we calculate the average difference over all data, we get the so-called empirical variance:

$$S^2 = \frac{1}{n} \sum_{i=1}^{n} (x_i - \bar{x})^2$$

The root of this expression is called empirical standard deviation: $S = \sqrt{S^2}$.

For our example we have the following variance (without the units for simplicity reasons):

$$S^2 = \frac{1}{7} ((-5)^2 + (-4)^2 + (-5)^2 + (-6)^2 + (-6)^2 + 30^2 + (-4)^2) \approx 150.57$$

And for the standard deviation:

$$S \approx 12.27$$

A deviation which is much larger than the average value is a sure sign for the fact that the average is dominate by a few values and does not really represent the data.

If we calculate the average for all building of the city without the church, we get: $\bar{x} = 5$.

The variance is:

$$S^2 = \frac{1}{6} (1^2 + 1^2 + 0^2 + 0^2 + (-1)^2 + (-1)^2) = \frac{2}{3}$$

The standard deviation is about $S = 0.82$ – which is significantly smaller than the standard deviation we had when we included the church, and significantly smaller than the average height of the houses. In this case, the average is a good representation for the height of the houses.

In addition to the average and the standard deviation, we can also use the median to describe our set of data. The median $\tilde{x}$ divides the set of data in two equal halves. In general: If our set of data is

$$x_1, x_2, \ldots x_{\frac{n+1}{2}}, \ldots x_n$$

then $x_{\frac{n+1}{2}}$ is the median. In our case (the numbers are sorted by size) we have:

$$4, 4, 5, 5, 6, 6, 40$$

Our set of data contains seven data, so the forth data is the median, i.e. $\tilde{x} = 5m$. This also gives us a better impression about the height of the buildings in the city than the average of 10m.

A measure like the median is also called a quantile. In the case of the median, 50% of all data are below a certain value. In the case of a 25%-quantile, 25% of all data are below this number. In the case of our buildings this would be the height of 4m.

In the case of a 75%-quantile, 75% of all data would be below this value. In our case this would be the height of 6m.

In statistics, we often ignore "outliers" that are too far away from the average. Usually, we only look at values between the 10%-quantile and the 90%-quantile. The 10% below and the 10% above this limit are treated as outliers that could distort the evaluation of the data.

Sometimes we do not have data of one kind like the height of buildings, but we measure two values that depend on each other like the velocity of a car and the gas consumption or the difference in air pressure and the speed of the wind. We would like to know how these values depend on each other. Often we are able to provide detailed formula, but they might depend on constants that are not always known. Statistics can help us to find a value for these constants.

In the simplest case the dependency between two measurement values is given by a linear equation of the form

$$f(x) = ax + b$$

The variables x and $f(x)$ are the related measurement values, and a and b are the unknown constants.

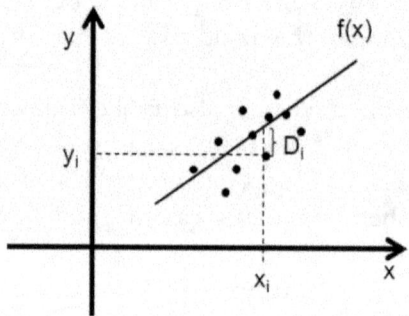

Figure 28: Linear regression: We are looking for the function ax+b which has on average the smallest distance to the measurement points.

We have a set of data which can be imagined as a "cloud of points" in a coordinate system. We are now looking for the straight line (or curve in general) that fits best to the data, i.e. which has the smallest distance to all measurement points. Figure 20 shows this for the example of a linear function.

We want to minimize the difference D_i to the function $f(x_i)$ for the pair of data (x_i, y_i). The difference is calculate as

$$D_i = y_i - f(x_i) = y_i - ax_i - b$$

Just like in the case of the standard deviation we could calculate with the absolute value, but using the square is easier. The method to calculate the best fit is thus called the "method of least squares".

As the squares over all measurement data have to be minimal, we have to calculate the minimum of the function

$$f(a, b) = \sum_{i=1}^{n} D_i^2 = \sum_{i=1}^{n} (y_i - ax_i - b)^2$$

In this case, the values x_i and y_i are given, so the function depends on the variables a and b.

The necessary condition for a minimum of a function is that the derivative with respect to the variable is zero. As we have two variables, the partial derivatives with respect to these two variables have to be zero, i.e. $f(a, b)$ has to meet the conditions

$$\frac{\partial f(a, b)}{\partial a} = 0$$

and

$$\frac{\partial f(a, b)}{\partial b} = 0$$

Before we can differentiate the summand, we multiply the square and get

$$f(x) = \sum_{i=1}^{n}(y_i^2 - 2ax_iy_i - 2by_i + a^2x_i^2 + 2abx_i + b^2)$$

The differentiations yield:

$$\frac{\partial f(a,b)}{\partial a} = \sum_{i=1}^{n}(-2x_iy_i + 2ax_i^2 + 2bx_i)$$

$$\frac{\partial f(a,b)}{\partial b} = \sum_{i=1}^{n}(-2y_i + 2ax_i + 2b)$$

These derivatives have to be zero. We divide by two and order the sums to get

$$\left(\sum_{i=1}^{n}x_i\right) \cdot b + \left(\sum_{i=1}^{n}x_i^2\right) \cdot a = \sum_{i=1}^{n}x_iy_i$$

$$n \cdot b + \left(\sum_{i=1}^{n}x_i\right) \cdot a = \sum_{i=1}^{n}y_i$$

In principle, we could solve this equation analytically. But as we normally have a large number of measurement data, the computer helps to simplify the calculation.

The method of least squares cannot only be used for linear function, but also for polynomials of any kind. A polynomial is an equation of the form:

$$f(x) = a_n x^n + a_{n-1} x^{n-1} + \ldots + a_1 x + a_0$$

If the relationship between the two data is unknown, we have to start our calculation with this general form of the solution for $f(x)$. The calculation will show which summands are really needed for the best approximation of the measurement data.

Such a regression can only provide a meaningful relationship if the set of data is big enough. With only two data points we could come up with any graph.

With statistics we are only approximating our measurement data. And as we have seen just calculating the average could give a completely wrong impression about our set of data. These examples show that statistics isn't such an exact science as adding up numbers. After all, we want to represent a large set of data by just a few numbers, which inevitably means that we have to ignore some information. This is nothing illegitimate. The MP3-code is doing just the same. It ignores this information in a piece of music that our clumsy and not so precise human ear doesn't hear anyway.

The important thing is to make sure we don't ignore the wrong information – and our piece of music only contains the bass line without the melody.

The same is true in statistics. If we ignore the wrong information, we can provide a wrong impression – and lie splendidly with statistics.

Before we start with a statistical analysis of our data, we should ask ourselves how reliable these data are. There is a reason that most prices in the supermarket end with 9 cents, like 0.99 or 1.99. Some say that the reason is that the customer gets the impression that the goods are cheaper because he would just look at the number before the point and ignore the 99 cents behind it. But frankly, who calculates with 1 Euro if he sees a price of 1.99? Everyone brings the price up to a round figure and calculates with 2 Euros.

But a price of 1.00 Euro or 2.00 Euros seems to be strange. Even more: It doesn't seem to be authentic. It gives the impression that someone just estimated the price without taking the pain of actually calculating it exactly. If we see even numbers we feel cheated. So we get odd numbers like 0.99.

This can be delicate in the case of measurement data. We want to give the impression that our data are exact and reliable – so we "improve" them. Even if our equipment only allows to measure speeds with a precision of 1 m/s, we feel better if we can present at least one position after the decimal point. Even if we can measure a length only with the precision of 1 cm, we feel better to add the millimeters. The more precise our measurement data the more precise our analysis will be...

Sometimes you might have exact measurement values, but you don't want the reader to see them because they do not really reflect your theory. Quite often we want to show a significant change in our analysis. "The drug reduces the probability of complications by one third." This is a statement we like to see. And it doesn't have to be a lie. But quite often the database is so thin that you cannot truthfully make any statement about a change – so you better avoid showing the basic data and only present the analysis.

Let's assume we have examined 1000 patients that were suffering from a certain disease. Three patients had complications when not using a certain drug, and only two patients had complications when using the drug. This means that only 0.3% of all patients had suffered from complications in the first case, 0.2% in the second case. This change is so small that it could be a pure accident. If we saw this numbers in an advert for a drug, we wouldn't think that it's worth taking the drug.

So the advert only tells us about the relative change. Using the drug, one patient out of three didn't have complications, so the drug reduced the complications by one third. That is much more impressive – and we are convinced that it would make sense to take this drug.

Mentioning relative changes is very popular, as they easily provide huge numbers. Some companies thus tell us that their growth of revenues increase by 20% from last year to this year. The revenues must have skyrocketed!

But again: We do not compare absolute values but relative values. The revenue-growth of the company might have been 1% last year; it is 1.2% this year. Bluntly, we would say it didn't change at all. But 1.2% is 20% more than just 1% - and so we get the information that the revenue growth increased by 20%. If you want to be more truthful, you could say that the revenue growth increased by 0.2 percent point – but again that's not really impressive.

You can also twist the truth in graphs, even if you are drawing the absolute numbers. It is only a question of the scale.

One example is shown in figure 29. The upper graph shows the change of revenue in the last ten years in absolute numbers. You can see that the revenue didn't change much in this period of time (it increased by only 5.3%).

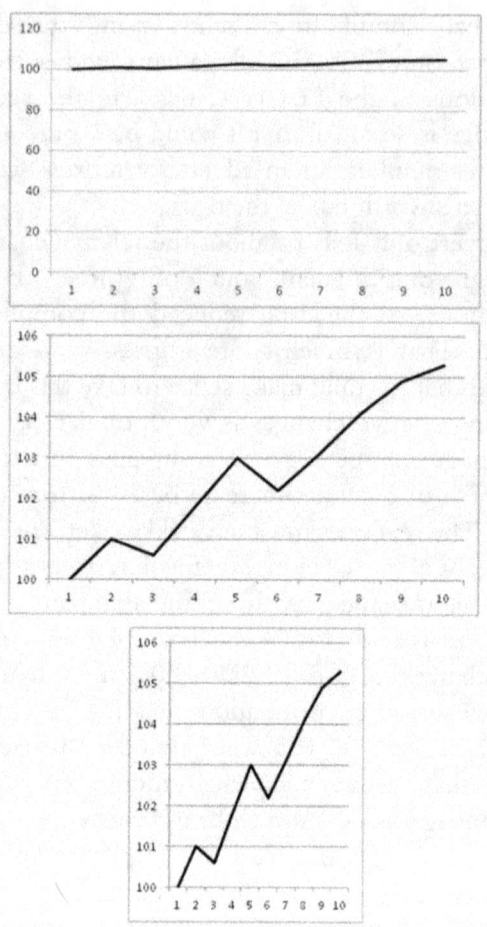

Figure 29: The change in revenue, scaled in three different ways.

But if we omit the point of origin and start our scale of the y-axis at 100, then we have a much more impressive development, as the second graph shows. Admittedly, there was a setback in the third and sixth year, but apart from that, the company grew astoundingly. The revenue growth is even more impressive if the x-axis is compressed, as can be seen in the third graph.

We have the same numbers in all three cases. But in the first graph we only see a very small growth, almost stagnation, while the company grew impressively in the last graph. It's just a question of how you present your data.

It is also very important to decide which data you want to show. The long-term trend might be that a situation improves. But if your theory predicts deterioration, you just have to choose the right period of time as figure 30 shows.

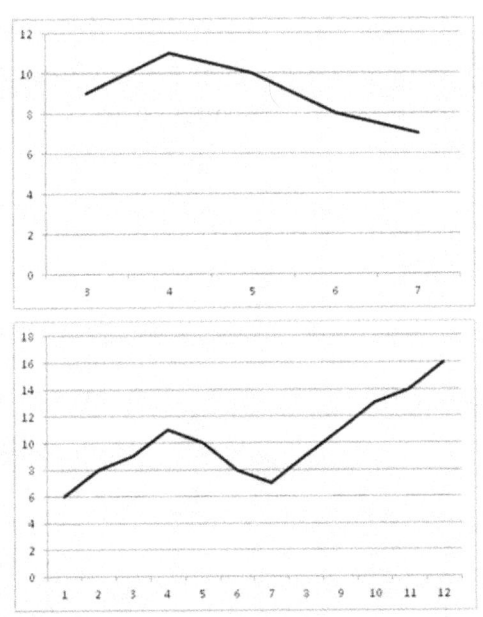

Figure 30: Growth or not? It only depends on the right period of time…

Looking at the upper graph we clearly see that everything gets worse. But looking at the overall picture in the second graph, we see that this was just a momentary weakness, the overall trend is clearly positive. Depending on what your statement is, you can choose the upper or the lower graph to support it. Lying with statistics is very easy, thus we have to be careful when looking at them.

5.2. Calculus of Probabilities

Statistics deals with the analysis and evaluation of data. The calculus of probabilities tries to calculate the behavior of random phenomena.

The set of all possible outcomes are called the elementary events ω_i. The set of all elementary events

$$\Omega = \{\omega_1, \omega_2, \omega_3, ...\}$$

is called the sample space. A subset of the sample space is called an event (e.g. A). The set of all events that can be created from the samples space Ω is called the event space E.

Let us take the example of throwing dice. The thrown numbers are the elementary events. The set of all numbers is the sample space and the set of all even numbers is an event.

For two events A and B we have the following definitions:
- The event $A \cap B$ happens when A and B happen (we have the intersecting set). If we have $A \cap B = \emptyset$, then the intersecting set is an empty set. The two events A and B exclude each other.
- The event $A \cup B$ happens if one of the events A or B happens. We have the set union of the two events.

- The symbol $\bar{A}$ designates the complementary event to A, i.e. it happens if A doesn't happen. We get if if we subtract the set A from the set Ω ($\Omega \backslash A$). $\bar{A}$ is the remaining or difference amount of A.

Now we ask ourselves: What is the probability that the event A happens?

To calculate this, we attach a probability to each elementary event, as this was first done by the French mathematician Pierre-Simon Laplace in the 18th century. If the set of elementary events has n elements, then each elementary event has the probability of $\frac{1}{n}$. If we sum up the probabilities of all elementary events we get the sum probability of 1. The definition of probability is such that it can only have values between 0 and 1.

The probability $P(A)$, that the event A happens, is defined as

$$P(A) = \frac{|A|}{n} = \frac{\text{number of elements of } A}{\text{number of elements of } \Omega}$$

Laplace introduced this definition. It has the consequence that the probability of the event A is higher the higher the number of elementary events is that can lead to this event.

Let us take the example of throwing a coin. We can have head (H) or tail (T). We measure this set of elementary events:

$$\Omega = \{H, T, T, H, H, T, T, T, H, H\}$$

The probability to get heads when throwing a coin is

$$P(H) = \frac{5}{10} = 0.5$$

In another experiment we could have found that we only have four times head, then the probability for this event would have been 0.4. Only when we have a very large set of elementary events we can be sure that the empirical probability for an event will be the same as the theoretical probability. To get the theoretical probability, we write down all possible combinations.

Let's, for example, assume that we are throwing a die and a coin. The die can produce the elementary events of 1 to 6. The question now is: What is the probability to get heads with the coin and an even number with the die?

To answer that question, we write down all possible combinations:

$$\Omega = \{(H,1),(H,2),(H,3),(H,4),(H,5),(H,6), \\ (T,1),(T,2),(T,3),(T,4),(T,5),(T,6)\}$$

The event A was defined as the combination of heads with an even number. The set is:

$$A = \{(H,2),(H,4),(H,6)\}$$

This gives a probability of:

$$P(A) = \frac{3}{12} = 0.25$$

That means the probability for such an event is 0.25 (or 25%).

The probability to have event A can depend on another event B. If event A can only happen if event B has happened, then we have a conditional probability, written as $P(A|B)$. The probability of $P(A|B)$ is equal to the probability of the intersecting set of A and B (both have to happen) divided by the probability that B happens (this probability is now the new frame of reference and no longer the probability 1, the overall set of all elementary events Ω). Thus we have:

$$P(A|B) = \frac{P(A \cap B)}{P(B)}$$

The analogous formula for the probability that B happens if the event A has happened is:

$$P(B|A) = \frac{P(A \cap B)}{P(A)}$$

This give as the multiplication law for probabilities:

$$P(A \cap B) = P(A) \cdot P(B|A) = P(B) \cdot P(A|B)$$

From this law, we instantly can derive Bayes' theorem, which is named after the British mathematician Thomas Bayes who lived in the 18th century:

$$P(A|B) = \frac{P(A) \cdot P(B|A)}{P(B)}$$

The remarkable consequence of this theorem is that it allows us to calculate the probability of the event A which depends on the event B, even if we only know the probability of the event B depending on the probability of the event A.

We can show this with a simple example. Let us assume that we have a disease which infects 10 out of 100,000 people. The appearance of the disease (A) has the probability $P(A) = 0.0001$.

But we are lucky. We have a test for this disease. Let B be the probability that the test gives a positive result (i.e. it tells us that someone is infected).

However, no test is perfect. Our test can detect the disease only with a probability of 98%. The probability is thus $P(B|A) = 0.98$. On the other hand, in one percent of the tests we get a positive result, even if the person is not infected, i.e. the test finds "ill" people in the set of healthy people $\bar{A}$ (in the set of those people that do not belong to the group of ill people A). The probability for this event $P(B|\bar{A})$ is therefore 0.01. The probability $P(\bar{A})$ that a person is not ill is 0.9999 (as we have $P(\bar{A}) = 1 - P(A)$).

We know the probability $P(B|A)$ that the test is positive if we have an infected person (it is 0.98). The question now is: What is the probability that someone is really ill if the test result is positive? The probability A that a person is really ill now depends on the positive result of the test (the event B). Wir are looking for the probability $P(A|B)$.

The Bayes' theorem was:

$$P(A|B) = \frac{P(A) \cdot P(B|A)}{P(B)}$$

$P(B)$ is the probability that we get a positive result with the test. We are now testing the group of infected people (A) and the group of healthy people ($\bar{A}$). The test can deliver positive results in both groups. The probability $P(B)$ that an ill person is tested as ill is $P(B|A) \cdot P(A)$ (the probability of a positive result multiplied with the probability to be ill).

The probability that a healthy person is tested as ill is $P(B|\bar{A}) \cdot P(\bar{A})$. So the overall probability $P(B)$ to get a positive test-result is:

$$P(B) = P(B|A) \cdot P(A) + P(B|\bar{A}) \cdot P(\bar{A})$$

If we put this into Bayes' theorem, then we have:

$$P(A|B) = \frac{P(A) \cdot P(B|A)}{P(B|A) \cdot P(A) + P(B|\bar{A}) \cdot P(\bar{A})}$$

If we use our numbers, we get:

$$P(A|B) = \frac{0.0001 \cdot 0.98}{0.98 \cdot 0.0001 + 0.01 \cdot 0.9999} \approx 0.001$$

The probability to be really ill if the test result is positive is only 0.001 or 0.1%.

That result is astounding. After all, the test promises to detect the disease with an exactitude of 98%, so how can it be that the probability to be ill is only 0.1% if we have a positive results? Did we make a mistake in our calculation?

We didn't. To understand this, we have to take a closer look at the different probabilities. We can use a tree diagram for this as shown in figure 31.

Let's assume we have tested one million people. If the probability for the disease is 10 of 100,000, then we can assume that we have 100 ill people and 999,900 healthy people in our test group. The test detects the disease with a probability of 98%, i.e. in the group of ill people we will really find 98 ill people. Two people are sorted out as healthy.

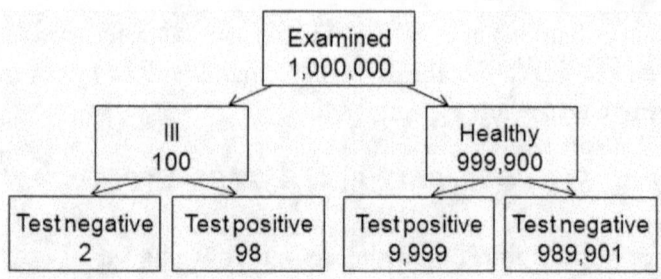

Figure 31: Tree diagram to answer the question: How many positively tested persons are really ill?

On the other hand, the test gives us a wrong positive result in one percent of the cases. This means that 9,999 people are erroneously declared ill by the test. This is a much higher number than in the group of ill people, because there are so much more healthy people. In sum, the test will find 10,097 ill people in a group of 1,000,000 people, but in fact only 98 of them are really ill, i.e. we have a quota of correctly identified ill people of

$$P(A|B) = \frac{98}{10,097} \approx 0.001 = 0.1\%$$

This is one of the reasons why some scientists are skeptical with regards to all the preventive medical checkups that we are asked to do.

*

It is not always the case that the event A depends on the event B. Both events can be independent from each other. Then we have:

$$P(A|B) = P(A)$$

And the multiplication law is simply:

$$P(A \cap B) = P(A) \cdot P(B)$$

Let's assume we would have a technical device made of 100 components, and the device is built in a way that the failure of one component leads to a complete failure of the device. On components has a probability of $p = 99\%$ to work. What is the probability that the whole device will work?

The device works if all of its components are working. The probability $P(A)$ that the device works is thus the intersecting set of the probabilities A_i that each components works (which is $p = 0.99$). The multiplication law gives:

$$P(A) = P(A_1 \cap A_2 \cap ... \cap A_{100})$$
$$= P(A_1) \cdot P(A_2) \cdot ... \cdot P(A_{100})$$

The probabilities $P(A_i)$ are all equal to p, and we get:

$$P(A) = 0.99^{100} \approx 0{,}37$$

The probability that the whole device is working is only 37%. Modern devices consist of more than 100 components. That's why manufacturers only accept failure rates of a few hundred ppm (parts per million) when they buy components for their devices.

Another case of independent events is the case of a box with white and black balls. The box contains N balls. M of these balls are black, and $N - M$ of these balls are white. The probability to pull out a black ball is $p = \frac{M}{N}$.

We are now randomly taking n balls out of the box, note their color and put them back. What is the probability to get k black balls?

We are pulling n balls out of the box and get k black balls and $(n-k)$ white balls. The probability to get a black ball is p, thus the probability to get a white ball is $(1-p)$. We call the elementary event to get a black ball A_i, and the elementary event to get a white ball $\bar{A}_i$. We are taking n balls out of the box, so the overall probability is:

$$P(A) = P(A_1 \cap ... \cap A_k \cap \bar{A}_{k+1} \cap ... \cap \bar{A}_n)$$

$$P(A) = P(A_1) \cdot ... \cdot P(A_k) \cdot P(\bar{A}_{k+1}) \cdot ... \cdot P(\bar{A}_n)$$

$$P(A) = p^k(1-p)^{n-k}$$

Now we also have to consider that the elementary events can have a different order. In the example above we first got all the black balls and then all the white balls. It can also be that we first get one black ball and then all white balls and then the remaining black balls. There are many different combinations of how we can pull out the balls. This number of combinations is given by the binomial coefficient

$$\binom{n}{k} = \frac{n!}{k!\,(n-k)!}$$

where $n! = n \cdot (n-1) \cdot ... \cdot 1$.

In the case of the balls we are dealing with a binomial distribution of the form

$$P(A) = \binom{n}{k} p^k (1-p)^{n-k}$$

If we have $N = 10$ balls and $M = 3$ are black, then we have $p = \frac{3}{10}$. If we pull $n = 5$ balls out of the box and put them back, then the probability to get $k = 2$ black balls is:

$$P(A) = \frac{5!}{2!\,3!}\left(\frac{3}{10}\right)^2 \left(\frac{7}{10}\right)^3 \approx 0.31$$

Our main interest might be the probability to win the lottery. In some lotteries, you need 6 out of 49 numbers to win the jackpot. But we cannot use the binomial distribution to calculate this probability as this one assumes that we put the balls back after we have noted the number. In a lottery the ball is not put back.

We have the situation that we start with $N = 49$ balls, and $M = 6$ of them are the "right" ones. We pull $n = 6$ balls out of the box and would like to know the probability to have k "right" ones (k can be any integer up to 6).

We combine the probabilities to take k right balls from the set of M balls and $(n - k)$ wrong balls from the set of $(N - M)$ wrong balls. Again, we can pull out the balls in arbitrary order, i.e. we have to use the binomial coefficient:

$$\binom{M}{k}\binom{N-M}{n-k}$$

This expression, however, is not complete. The definition of probability requires that it can only have values between 0 and 1. The expression above, however, can be infinitely large. To normalize it, we have to divide the expression by the overall probability to pull out n balls from a set of N balls. The order is arbitrary again, so this is also describe by the binomial coefficient $\binom{N}{n}$. The probability to win in a lottery is thus given by:

$$P = \frac{\binom{M}{k}\binom{N-M}{n-k}}{\binom{N}{n}}$$

In the special case where you can pull six balls out of a box which contains 49 balls, we have:

$$P = \frac{\binom{6}{k}\binom{43}{6-k}}{\binom{49}{6}}$$

The probability to have $k = 3$ right numbers is about 1.76%. The probability to have six right numbers is only 0.000007%. To put it differently: Only one out of 14 million players can hope to have the 6 right numbers and win the jackpot.

*

Sometimes we like to know the probability that a variable A has a value which is lower or equal to a given number x. We define the probability or distribution function to calculate this as:

$$F(x) = P(A \leq x)$$

Let's take the example of a die. The variable A can have values from $x_1 = 1$ to $x_6 = 6$. The probability $P(x_i)$ is $\frac{1}{6}$ for all values. The probability to get the number 1 to 5 is equal to the sum of the elementary probabilities, i.e. we have $F(5) = \frac{5}{6}$. Figure 32 shows the function $F(x)$.

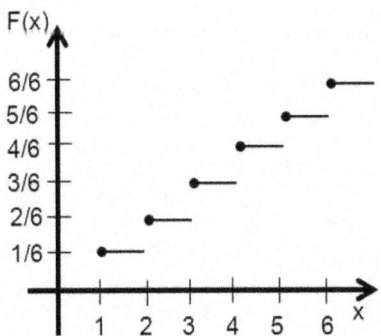

Figure 32: The distribution function F(x) for the case of a die.

Quite often, the set of data is less clearly arranged as in the case of a die. Sometimes, the variable A can have any value, so that the distribution function is not a step function but a continuous function. In this case, we cannot sum up the probabilities of the elementary events, but have to replace the summation by an integral. The behavior of the probabilities is given by the probability density function $f(x)$ and we have:

$$F(x) = P(A \leq x) = \int_{-\infty}^{x} f(t)dt$$

Obviously, we have for the distribution function:

$$\lim_{x \to -\infty} F(x) = 0$$

$$\lim_{x \to +\infty} F(x) = 1$$

The integration from $-\infty$ to $-\infty$ is zero (we have no event). And as the distribution function shall describe the overall probability of the system, it cannot be bigger than 1 per definition (then we have a sure event).

For an equal distribution of elementary events in the interval $[a, b]$ we can write the density function as:

$$f(x) = \begin{cases} \dfrac{1}{b-a}, & \text{for } a \leq x \leq b \\ 0, & \text{else} \end{cases}$$

We get the distribution function if we integrate the density function in the interval [a, b]:

$$F(x) = \int_{-\infty}^{x} f(t)dt = \int_{a}^{x} \frac{1}{b-a} dt = \frac{t}{b-a}\bigg|_{a}^{x} = \frac{x-a}{b-a}$$

This gives the following distribution function from 0 to infinity:

$$F(x) = \begin{cases} 0, & \text{for } x < a \\ \dfrac{x-a}{b-a}, & \text{for } a \leq x \leq b \\ 1, & \text{for } x > b \end{cases}$$

Figure 33 shows the density function and the distribution function for an equal distribution of probabilities.

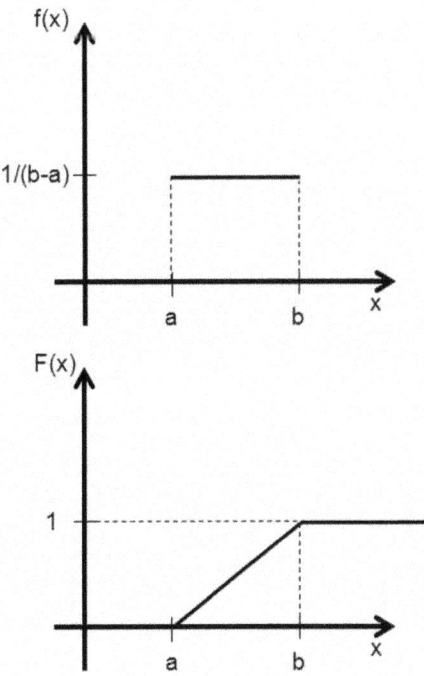

Figure 33: The density function f(x) and the distribution function F(x) for the equal distribution on the interval [a, b].

In statistics we defined the average as the sum over all values x_i divided by the number of values n. Using a similar approach, we define the expected value as the integral over all values multiplied with their probability (i.e. the density function). For the variable A the expected value $E(A)$ is therefore:

$$E(A) = \int_{-\infty}^{+\infty} x\, f(x)\, dx$$

Similarly, the variance σ^2 (σ is the Greek letter sigma) is given by

$$\sigma_A^2 = \int_{-\infty}^{+\infty} (x - E(A))^2 f(x)dx$$

The standard deviation is then defined by $\sigma_A = \sqrt{\sigma_A^2}$.
For the example of the equal distribution we get the expected value

$$E(A) = \frac{1}{b-a}\int_a^b x\,dx = \frac{1}{b-a}\frac{x^2}{2}\bigg|_a^b = \frac{b^2 - a^2}{2(b-a)}$$
$$= \frac{a+b}{2}$$

The "average" of the distribution is, as expected, in the middle of the interval $[a, b]$. The variance is calculated to

$$\sigma_A^2 = \frac{(b-a)^2}{12}$$

As we have mentioned the binomial distribution

$$P(A) = \binom{n}{k} p^k (1-p)^{n-k}$$

quite often, we should just note its expected value and variance:

$$E(A) = np$$

$$\sigma_A^2 = np(1-p)$$

Another important distribution is the normal distribution or Gaussian distribution, named after the German mathematician Carl Friedrich Gauss who introduced it in the 19th century. It is probably one of the most often used distributions, as many experimental data in science and engineering can be approximated very well with it. In addition, it is often used to calculated probabilities in many areas, e.g. in the insurance industry. The density function of the normal distribution seems to be quite complicated (φ is the Greek letter phi):

$$\varphi(x, \mu, \sigma) = \frac{1}{\sqrt{2\pi\sigma^2}} e^{-\frac{(x-\mu)^2}{2\sigma^2}}$$

But this notations allows us to get the variance σ^2 and the expected value $E(A)$ quite easily, because the expected value is nothing but μ. As figure 34 shows, the Gaussian distribution is symmetric with respect to the expected value, so that the maximum of the graph is the expected value. The variance or the standard deviation σ describes the width of the distribution.

We have approximately 68% of all values in the interval $\pm\sigma$ around the expected value. We cover 99.7% of all values in the interval of $\pm 3\sigma$ around the expected value.

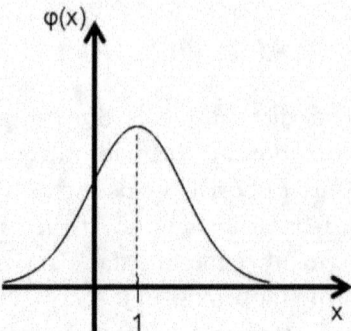

Figure 34: The Gaussian distribution for μ=1 and σ²=2.

The function φ is also called bell-shaped curve. The distribution function

$$\Phi(x,\mu,\sigma) = \int_{-\infty}^{x} \varphi(t,\mu,\sigma)\,dt$$

(with the large letter phi) is called Gaussian integral. It cannot be calculated analytically.

At the end of this paragraph, we will have a look at a strange distribution that cannot really be explained. It can only be said that it proved to be valid in many cases. It is the Benford distribution, also called Benford's law. It was discovered empirically by the American physicist Frank Benford in the middle of the 19th century (in fact it had already been discovered by the British mathematician Simon Newcomb at the end of the 18th century, but it was forgotten as it sounded too strange).

Both had discovered that numbers in large set of numbers are not distributed equally. Newcomb had observed that the pages in books with logarithm tables where the first numeral was a one were dirtier than all other pages. Obviously they had been used much more often.

Later on it was seen that the area of lakes, the population figures, share prices or tax data are not equally distributed either but show a strange clustering: Values with small first numerals appeared more often that values with large first numerals. The empirically found probability for the first numeral is:

$$h(1.\ \text{numeral of the number i}) = log_{10}\left(1 + \frac{1}{i}\right)$$

Figure 35 shows the typical distribution of the numerals 1 to 9. And even though no-one can explain why we have this distribution, it is used in many cases.

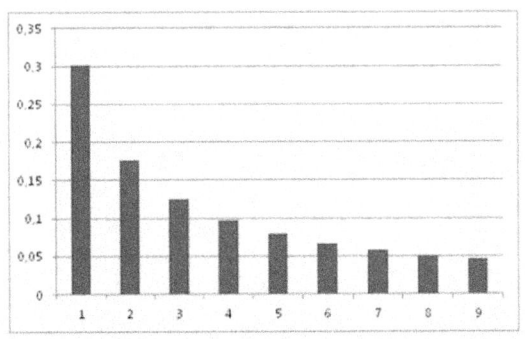

Figure 35: Benford's law: Numbers that start with a small numeral appear more often than numbers with a large numeral.

Benford's law is commonly used to check large data sets for manipulation. It was, for instance, used to discover the "creative" book-keeping of the bankrupt companies Enron and Worldcom. Fiscal authorities in Germany use this law since the beginning of the 21st century to discover tax fraud.

And even in science this law is used to detect manipulated data.

It doesn't necessarily mean that something is wrong when Benford's law is violated. Many distributions are normal distributions and don't follow Benford's law. But if this law is violated in economics, it is generally worth to take a closer look.

6. Analytic Geometry

6.1. Euclidean and non-Euclidean Geometry

For centuries no-one questioned the rules of geometry. The basic postulates were written down by Euclid already in the 3rd century B.C., and they are still the rules of geometry that we learn at school.
Euclid had written down five simple postulates on which he had based the whole science of geometry. These five postulates are:

1. It is possible to draw a straight line from any point to any point.
2. It is possible to extend a line segment indefinitely in a straight line.
3. It is possible to describe a circle with any center and radius.
4. We have that all right angles equal one another.
5. We have that, if a straight line falling on two straight lines makes the interior angles on the same side less than two right angles, the two straight lines, if produced indefinitely, meet on that side on which are the angles less than the two right angles.

The fifth postulate is the so-called parallel postulate. It can also be said that to a straight line we can only find one second straight line through a given point outside of the first straight line so that both lines are parallel.

The fifth postulate sounds rather awkward and complicated, especially when compared to the first four postulates. Mathematicians have tried to derive the fifth postulate from the first four, but they never succeeded.

Base on Euclid's geometry, Descartes developed his coordinate system where the position of a point is defined by its distance to two axes in a plane. In a three dimensional world, the position is defined relative to the three axes of the volume.

Sometimes there are problems with a rotational symmetry where the description of a point as the distance to some axes is not very intuitive. It would be much simpler to describe the position of the point with its distance r from the point of origin and the angle φ relative to one axis. Such "polar coordinates" were introduced a few years after the development of the Cartesian coordinate system. Figure 36 shows a coordinate system with polar coordinates.

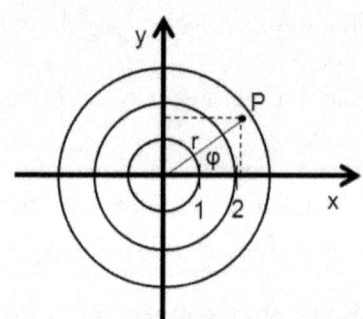

Figure 36: Polar coordinates: The Cartesian coordinates (x, y) are replaced by the distance r and the angle φ relative to the x-axis.

We can see that the coordinates form a right-angled triangle. If a point has the coordinates (x, y) in the

Cartesian coordinate system and the coordinates (r, φ) in the polar coordinate system, then we have the relationship:

$$x = r \cos \varphi$$
$$y = r \sin \varphi$$

We can expand the two dimensional Cartesian coordinate system by adding a third axis (the z-axis) which is perpendicular to the x-y-plane. We can also expand the polar coordinate system by adding a third axis. We then have:

$$x = r \cos \varphi$$
$$y = r \sin \varphi$$
$$z = z$$

These coordinates are called cylindrical coordinates, as they can best be used to describe problems with a cylindrical symmetry, as figure 37 shows.

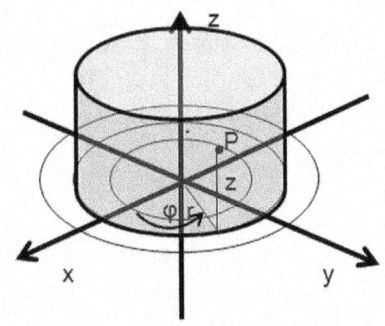

Figure 37: Cylindrical coordinates are polar coordinates with an additional z-axis.

If we have spherical symmetry, then we use spherical coordinates as shown in figure 38. The point P is defined by its distance r from the point of origin, the angle φ between the imaginary projection of the line r onto the x-y-plane and the x-axis, and the angle θ between the z-axis and the line r. The definition of the angle is so that $\theta = 0$ if the point P lies on the z-axis.

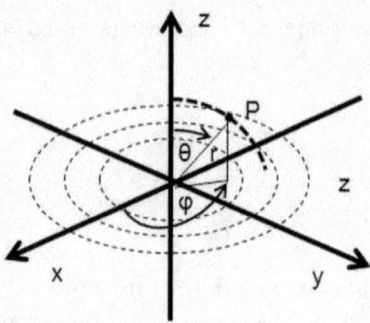

Figure 38: Spherical coordinates: The angle θ will be zero if P is on the z-axis.

The formulas to get from spherical to Cartesian coordinates are a little bit more complicated, but they can also be derived using the trigonometric relations in the right-angled triangle:

$$x = r \sin \theta \cos \varphi$$
$$y = r \sin \theta \sin \varphi$$
$$z = r \cos \theta$$

After our short review of coordinate system in the Euclidean geometry we come back to the parallel postulate. In the 19th century, mathematicians had shown that it was not possible to derive this postulate from simpler postulates. As mathematicians are curious people, some of them investigated what would happen to geometry if the postulate was wrong, i.e. if we either have

1. no parallel through a given point outside of a straight line, or if we have
2. several parallels through a given point outside of a straight line.

In the first case, we get something that is called elliptical geometry, in the second case we get something that is called hyperbolic geometry. Both are also called non-Euclidean geometries.

What are the consequences if we don't get a parallel line through a given point outside of a straight line?

Another way to put this is that both lines will intersect eventually. This may sound strange, but it is a quite common phenomenon. Just think of the meridians on Earth: They will intersect at the poles, even though they were parallel at the equator.

Figure 39 shows this for the surface of a sphere. Two meridians are shown for the Northern sphere of the Earth. They are perpendicular to the equator, i.e. they run parallel at this point. However, they intersect at the poles. This figure shows one of the consequences of the elliptical geometry: The rules known from a plane do not have to be valid anymore. The sum of the angles in a triangle in the plane is 180°. In a sphere, we already have two right angles at the equator and then another angle at the poles. So the sum of all angles is greater than 180° - and its precise value depends on the triangle.

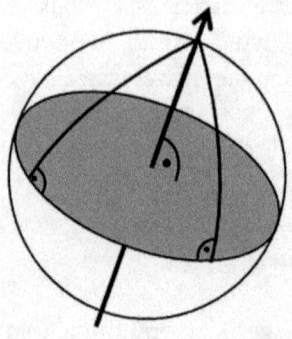

Figure 39: Elliptical geometry on a sphere: Meridians are parallel at the equator but intersect at the poles. The sum over all angles in the triangle is larger than 180°.

In hyperbolic geometry we have the case that we can find at least two parallel lines trough a given point outside of a straight line. To achieve this, our space needs to be bended in a way that we can have several straight lines through a point so that they are parallel to a first straight line and do not intersect anywhere else.

As a reminder: We say that a first straight line is parallel to a second straight line through a point P, if we have a line that is perpendicular to the first line and also perpendicular to the second line in the point P, just like the equator is perpendicular to two meridians.

In the case of the elliptical geometry the space was bended in a way that it moved together so that the straight lines also moved together and finally intersected.

In the case of the hyperbolic geometry we thus need a space that "moves apart". One example of such a space is a saddle as shown in figure 40.

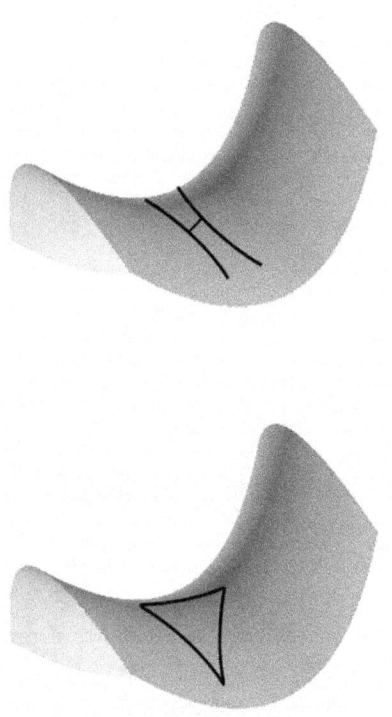

Figure 40: Hyperbolic geometry on a saddle: There is an infinite number of parallels to a straight line and the sum over all angles in a triangle is smaller than 180°.

If we draw a triangle on a saddle, we see that the sum of the angle in this triangle is smaller than 180°.
Now we have three different kinds of geometry, and we can calculate with all of them. The question is: Which of them actually describes our universe? The answer to this question is still unknown. It depends on the amount of mass that our

universe has. If there is so much mass in the universe that its expansion will stop one day and the universes contracts again, then the elliptical geometry would be the correct one. If the mass has just the right amount that the expansion will slow down and end one day, then the Euclidean geometry would be the right one. And if the mass is so small that the expansion will go on forever, we have to use the hyperbolic geometry.
But in mathematics such practical questions are not really important.

6.2. Vectors

We now return to the field of Euclidean geometry, a space where we can describe our objects with simple Cartesian coordinates. In this space we can define mathematical objects that we call vectors.

Up to now we have only dealt with numbers: How to calculate with them, how to establish a relationship between them using a function, how to integrate and differentiate a function. We can find numbers in our real world if we combine them with units like meter (the building has a height of 10m), kilograms (this person weights 80kg) or speed (the speed of this car is 100 km/h). Mathematical objects that are solely characterized by its value are called scalars.

But there can be other objects. We can understand this if we take a look at our last example, speed. The information how fast we are is important. But generally we also would like to know in which direction we are driving. So in addition to the pure value of speed we are also interested in the direction of speed and call it velocity. Mathematical objects that are defined by its value and the direction are called vectors.

To differentiate between a vector and a scalar we add an arrow on top of the variable. The vector a is thus written as $\vec{a}$. Unfortunately, this is not the nomenclature used in all books. Vectors can also be designated by a bold type (i.e. **a** for the vector a, as we will do this in the figures) or a vector may be designated by a line below the variable ($\underline{a}$). In this text we will use the arrow to designate a vector.

The value of the vector $\vec{a}$ is written as $a = |\vec{a}|$. To describe the direction of the vector, we use Cartesian coordinates. As we are only interested in the direction, we are only interested in the difference between the endpoint of the vector (shown by the arrow) and its beginning. We can say that a vector moves the start point by x-values along the x-axis and y-values along the y-axis to its endpoint. The vectors shown in figure 41 are thus identical even though they have different positions in the coordinate system.

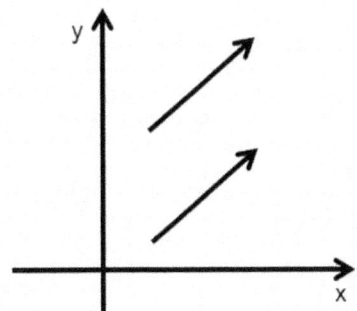

Figure 41: Both vectors show in the same direction and have the same length. They are therefore identical.

A vector in a plane is described with a pair of numbers. A vector $\vec{a}$ that moves an object by 2 in the x-direction and by 4 in the y-direction is written as:

$$\vec{a} = \begin{pmatrix} 2 \\ 4 \end{pmatrix}$$

A vector in three dimensions would be:

$$\vec{a} = \begin{pmatrix} x \\ y \\ z \end{pmatrix}$$

We can multiply a vector with a scalar. This is called scalar multiplication. The multiplication with a scalar just means that the vector becomes "longer"; it shows a greater change of direction in all dimensions. Multiplying a vector $\vec{a}$ with the scalar λ thus is nothing else but multiplying each component of the vector with the scalar:

$$\lambda \vec{a} = \begin{pmatrix} \lambda x \\ \lambda y \\ \lambda z \end{pmatrix}$$

The addition of two vectors is nothing but moving an object by first using the vector $\vec{a}$ and then adding a second movement described by the vector $\vec{b}$. The vector $\vec{b}$ starts at the end of vector $\vec{a}$. In sum, we could have started at the beginning of vector $\vec{a}$ and gone to the end of vector $\vec{b}$ directly to get the sum vector $\vec{c}$ as figure 42 shows.

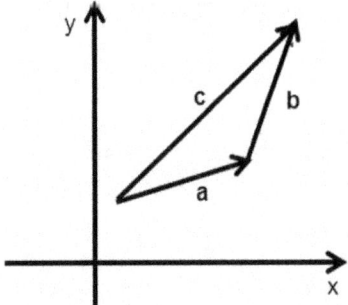

Figure 42: Addition of two vectors: a + b = c.

We get the sum vector $\vec{c}$ by adding the components of the original vectors. The addition

$$\vec{c} = \vec{a} + \vec{b}$$

thus means for the respective components of the vectors:

$$\begin{pmatrix} c_1 \\ c_2 \\ c_3 \end{pmatrix} = \begin{pmatrix} a_1 \\ a_2 \\ a_3 \end{pmatrix} + \begin{pmatrix} b_1 \\ b_2 \\ b_3 \end{pmatrix} = \begin{pmatrix} a_1 + b_1 \\ a_2 + b_2 \\ a_3 + b_3 \end{pmatrix}$$

The commutative law is valid for the components (as they are scalars), so it has to be valid for the vector as well:

$$\vec{a} + \vec{b} = \vec{b} + \vec{a}$$

The subtraction of two vectors $\vec{c} = \vec{a} - \vec{b}$ can be understood as the addition of the inverse vector $-\vec{b}$ to $\vec{a}$, i.e.

$$\begin{pmatrix} c_1 \\ c_2 \\ c_3 \end{pmatrix} = \begin{pmatrix} a_1 \\ a_2 \\ a_3 \end{pmatrix} + \begin{pmatrix} -b_1 \\ -b_2 \\ -b_3 \end{pmatrix}$$

So we have the commutative law for vectors, we have the inverse element, and we also the zero vector which is simply defined as

$$\vec{0} = \begin{pmatrix} 0 \\ 0 \\ 0 \end{pmatrix}$$

We have defined the length or the value of a vector as $a = |\vec{a}|$. But how do we calculate it for actual vectors?

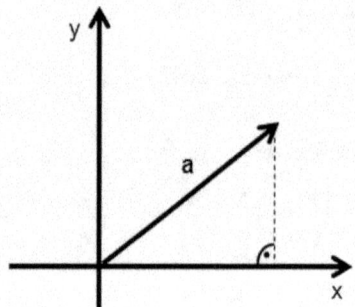

Figure 43: The length of a vector a can be calculated using the Pythagorean Theorem.

We can answer that questions easily if we move the vector into the point of origin (this doesn't change anything with respect to its value or direction). This is shown in figure 43. The coordinates x and y of the vector $\vec{a}$ form a right-angled triangle with the vector, thus we calculate its length with the Pythagorean Theorem:

$$a = |\vec{a}| = \sqrt{x^2 + y^2}$$

If the vector has n dimensions, we have

$$a = |\vec{a}| = \sqrt{a_1^2 + a_2^2 + \cdots + a_n^2}$$

If we add one additional axis to a two dimensional coordinate system, the axis is perpendicular to the original plane. If we add a fourth dimension, the forth axis is perpendicular to the three existing axes. You can't really imagine what this looks like – but you can calculate with it. So we don't have a problem to calculate the sum

$$\begin{pmatrix} 2 \\ 3 \\ 4 \\ 3 \end{pmatrix} + \begin{pmatrix} 4 \\ 3 \\ 2 \\ 5 \end{pmatrix} = \begin{pmatrix} 6 \\ 6 \\ 6 \\ 8 \end{pmatrix}$$

whatever this vector might look like.

Now we might ask how the third basic arithmetic operation, the multiplication, is defined for vectors. What can be the result if we multiply a vector with a vector?

A vector describes a direction. So we could expect that the product of two vectors will be a vector again, i.e. showing in a certain direction. On the other hand, a vector also has a certain length. The multiplication of two vectors could therefore result in a number, in a scalar.

And in fact, mathematicians have defined both versions of multiplications for vectors. Let's start with the inner product or scalar product. This is the multiplication that yields a numbers (and should be differentiated from the scalar multiplication which is the multiplication of a vector with a scalar).

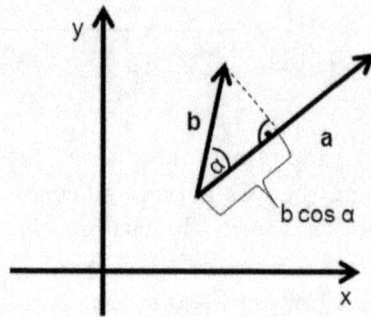

Figure 44: The inner product of the vectors **a** and **b** is equal to the product of the length of the vector **a** with the length of the projection of the vector **b** on **a**.

Our starting point are the two vectors $\vec{a}$ and $\vec{b}$ as shown in figure 44. The length of the vector $\vec{a}$ is a. The length of the vector $\vec{b}$ is b. Now we could simply assume that we multiply the lengths of the two vectors to get the inner product

$$c = |\vec{a}| \cdot |\vec{b}|$$

The situation, however, is not that easy as the vector $\vec{b}$ shows into another direction as the vector $\vec{a}$. The length of $\vec{b}$ that is "seen" by $\vec{a}$ is only the projection of the vector $\vec{b}$ onto $\vec{a}$. If α is the angle between the vectors $\vec{a}$ and $\vec{b}$, then the projection has the length $b \cos \alpha$. Thus, the inner product is:

$$\vec{a} \cdot \vec{b} = |\vec{a}| \cdot |\vec{b}| \cdot \cos\alpha = ab\cos\alpha$$

It is, however, not always easy to calculated the angle between two vectors. If we know the two coordinates of a two dimensional vector, then we can calculate its length using the Pythagorean Theorem. But calculating the angle is much more complicated and thus calculating the inner product in this way is usually not possible. But we can use this definition to derive a formula that allows us to calculate the inner product using only the coordinates of a vector.

Before we do this, we have to discuss shortly the unit vectors. In physics we have the meter as the unit of length. The length was originally defined with the help of a bar made from platinum-iridium which contained two notches in the distance of a meter. If you wanted to measures a length anywhere on the world you had to compare this length with the meter of this so-called standard meter.

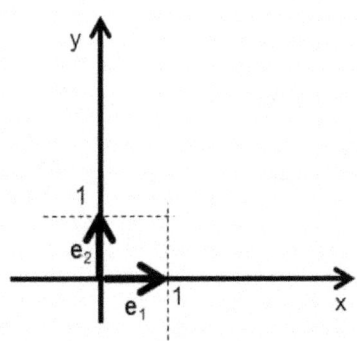

Figure 45: The unit vectors in two dimensions.

Likewise, the length of a vector has to be related to some sort of unit (like numbers use the number one as a unit). This unit is called the unit vector. A unit vector shows into the same directions as an axis in the Cartesian coordinate system and has a length of one. Figure 45 shows the unit vectors $\vec{e_1}$ and $\vec{e_2}$ for two dimensions. As can be easily seen, they have the form:

$$\vec{e_1} = \begin{pmatrix} 1 \\ 0 \end{pmatrix}; \vec{e_2} = \begin{pmatrix} 0 \\ 1 \end{pmatrix}$$

The unit vectors are perpendicular to each other, per definition. This means that we have $\cos\alpha = 0$. So the inner products are $\vec{e_1}\cdot\vec{e_1} = \vec{e_2}\cdot\vec{e_2} = 1$ and $\vec{e_1}\cdot\vec{e_2} = \vec{e_2}\cdot\vec{e_1} = 0$. Let us now take two arbitrary vectors

$$\vec{a} = \begin{pmatrix} a_1 \\ a_2 \end{pmatrix} = a_1\vec{e_1} + a_2\vec{e_2}$$

$$\vec{b} = \begin{pmatrix} b_1 \\ b_2 \end{pmatrix} = b_1\vec{e_1} + b_2\vec{e_2}$$

And calculate their inner product

$$\vec{a}\cdot\vec{b} = (a_1\vec{e_1} + a_2\vec{e_2})\cdot(b_1\vec{e_1} + b_2\vec{e_2})$$

$$= a_1\vec{e_1}b_1\vec{e_1} + a_2\vec{e_2}b_1\vec{e_1} + a_1\vec{e_1}b_2\vec{e_2} + a_2\vec{e_2}b_2\vec{e_2}$$

We remember that we have $\vec{e_1}\cdot\vec{e_1} = \vec{e_2}\cdot\vec{e_2} = 1$ and $\vec{e_1}\cdot\vec{e_2} = \vec{e_2}\cdot\vec{e_1} = 0$. So we get

$$\vec{a}\cdot\vec{b} = a_1b_1 + a_2b_2$$

To calculate the inner product of two vectors we simply multiply the coordinates with the same index and add the respective products. We can see that the commutative law is also valid for the inner product:

$$\vec{a} \cdot \vec{b} = \vec{b} \cdot \vec{a}$$

With this we are now able to calculate the angle between two vectors in an easy way:

$$\cos \alpha = \frac{\vec{a} \cdot \vec{b}}{|\vec{a}| \cdot |\vec{b}|}$$

Notably, we have the following rules for the inner product of vectors:

- If $\vec{a}$ and $\vec{b}$ are parallel and oriented in the same direction ($\alpha = 0$), then we have: $\vec{a} \cdot \vec{b} = ab$
- If $\vec{a}$ and $\vec{b}$ are parallel and oriented in the inverse direction ($\alpha = 180°$), then we have: $\vec{a} \cdot \vec{b} = -ab$
- If $\vec{a}$ and $\vec{b}$ are perpendicular to each other ($\alpha = 90°$), then we have: $\vec{a} \cdot \vec{b} = 0$

These rules can be derived directly using the definition of the cosine.

With the inner product we can quickly assess if two vectors are perpendicular to each other or, as mathematicians say, if they are orthogonal. For in this case, the inner product is zero.

Orthogonality is an important catchword for the second way to multiply vectors. This is called vector product or, because of the operator used, cross product.

The definition of the vector product is based on physical phenomena. The best known is probably the Lorentz force.

This force acts on a charged particle that moves through a magnetic field. The velocity of the particle defines the first vector, the orientation of the magnetic field the second vector. The Lorentz force now acts perpendicular to the velocity of the particle and the magnetic field, as can be seen in figure 46. A positively charged particle is deflected downwards if the particle moves from left to right, and the magnetic field is oriented out of the plane towards the viewer. We can remember the direction of deflection with the "three-finger-rule". The thumb shows into the direction of the first vector, the velocity. The index finger shows in the direction of the second vector, the orientation of the field. Then the middle finger automatically shows into the direction of the force (a negatively charged particle would be deflected in the inverse direction).

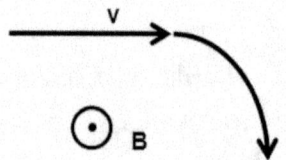

Figure 46: The Lorentz force: A magnetic field which shows into the direction of the viewer, deflects a positively charged particle, which moves to the right, downwards.

This observation inspired the definition of the vector product in mathematics. The vector product of the two vectors $\vec{a}$ and $\vec{b}$ results in a vector which is orthogonal to both vectors (i.e. orthogonal to the plane that this two vectors form). Its length is defined as the area of the parallelogram formed by the two vectors $\vec{a}$ and $\vec{b}$. And the three vectors form a right-handed system as described by the "three-finger-rule". Figure 47 shows the definition.

The area of a parallelogram with the sides a and b is equal to the product of the side a with the height of the parallelogram. We have for the height: $h = b \sin \alpha$, if α is the angle between the sides a and b. With $\vec{n}$ as the unit vector which is perpendicular to the plane described by $\vec{a}$ and $\vec{b}$, the vector product is defined as

$$\vec{a} \times \vec{b} = (|\vec{a}| \cdot |\vec{b}| \sin \alpha) \vec{n}$$

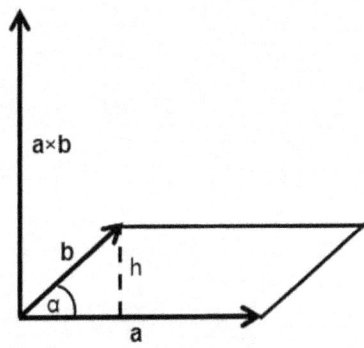

Figure 47: The vector product of the vectors **a** and **b** is a vector which is perpendicular to the plane of **a** and **b** and whose length is equal to the area of the parallelogram defined by **a** and **b**.

Again we have an angle α whose value is not clearly defined and easily calculable. In addition, we have the unit vector $\vec{n}$ that may have any value. This general definition doesn't really help to calculate the vector product of two given vectors. But we can again derive a simple formula to calculate the vector product with the coordinates of the vectors.

To do this, we take a look at how the unit vectors $\vec{e_1}, \vec{e_2}$ and $\vec{e_3}$ of the three axes behave in the case of the vector product. According to the three-finger rule we have:

$$\vec{e_1} \times \vec{e_2} = \vec{e_3}$$
$$\vec{e_2} \times \vec{e_3} = \vec{e_1}$$
$$\vec{e_3} \times \vec{e_1} = \vec{e_2}$$

And also (if we take the inverse direction through the axes):

$$\vec{e_2} \times \vec{e_1} = -\vec{e_3}$$
$$\vec{e_3} \times \vec{e_2} = -\vec{e_1}$$
$$\vec{e_1} \times \vec{e_3} = -\vec{e_2}$$

If we form the vector product of a unit vector with itself, then we get the zero vector as the angle between them and thus the sine are zero, i.e. we have:

$$\vec{e_1} \times \vec{e_1} = \vec{e_2} \times \vec{e_2} = \vec{e_3} \times \vec{e_3} = \vec{0}$$

Now we take two arbitrary vectors

$$\vec{a} = \begin{pmatrix} a_1 \\ a_2 \\ a_2 \end{pmatrix} = a_1 \vec{e_1} + a_2 \vec{e_2} + a_3 \vec{e_3}$$

$$\vec{b} = \begin{pmatrix} b_1 \\ b_2 \\ b_3 \end{pmatrix} = b_1 \vec{e_1} + b_2 \vec{e_2} + b_3 \vec{e_3}$$

and calculate the vector product:

$$\vec{a} \times \vec{b} = (a_1\vec{e_1} + a_2\vec{e_2} + a_3\vec{e_3}) \times (b_1\vec{e_1} + b_2\vec{e_2} + b_3\vec{e_3})$$

$$= a_1\vec{e_1} \times b_1\vec{e_1} + a_2\vec{e_2} \times b_1\vec{e_1} + a_3\vec{e_3} \times b_1\vec{e_1}$$
$$+ a_1\vec{e_1} \times b_2\vec{e_2} + a_2\vec{e_2} \times b_2\vec{e_2} + a_3\vec{e_3} \times b_2\vec{e_2}$$
$$+ a_1\vec{e_1} \times b_3\vec{e_3} + a_2\vec{e_2} \times b_3\vec{e_3} + a_3\vec{e_3} \times b_3\vec{e_3}$$

We can simplify this confusing expression if we calculate the vectors products of the unit vectors with the rules that we have just derived:

$$\vec{a} \times \vec{b} = \vec{0} - a_2 b_1 \vec{e_3} + a_3 b_1 \vec{e_2}$$
$$+ a_1 b_2 \vec{e_3} + \vec{0} - a_3 b_2 \vec{e_1}$$
$$- a_1 b_3 \vec{e_2} + a_2 b_3 \vec{e_1} + \vec{0}$$

If we order this expression with respect to the unit vectors, we get:

$$\vec{a} \times \vec{b} = (a_2 b_3 - a_3 b_2)\vec{e_1} + (a_3 b_1 - a_1 b_3)\vec{e_2} + (a_1 b_2 - a_2 b_1)\vec{e_3}$$

or

$$\vec{a} \times \vec{b} = \begin{pmatrix} a_2 b_3 - a_3 b_2 \\ a_3 b_1 - a_1 b_3 \\ a_1 b_2 - a_2 b_1 \end{pmatrix}$$

We can thus calculate the vector product in a simple way. The rows of the vector product are formed by the components of the two other rows. You start with the product of the next component of the first vector and the next but one component of the second vector. The second component is derived from the third row of the first and the fourth row of the second vector. As the vector only has three rows, you just start at the beginning, and the fourth row becomes the first row.

In the case of the inner product we had seen that the inner product of two vectors is zero if the two vectors are perpendicular to each other. This relationship is bijective, as mathematicians say, i.e. we can say that the inner product is zero when the vectors are perpendicular, and we can also say that the vectors are perpendicular if the inner product is zero. The statements "an inner product of zero" and "being perpendicular" are perfectly interchangeable. If one of the statements is true, the other one has to be true as well.

But what is the case if we have $\vec{a} \times \vec{b} = \vec{0}$?

This expression can be zero if the angle between the two vectors is zero, i.e. we know that the vectors are either parallel or anti-parallel, but we cannot say which of these two possibilities is true.

The commutative law was valid for the inner product, but it is not valid for the vector product. The vectors $\vec{a}$ and $\vec{b}$ are non-commutative (as they have to fulfill the "three-finger rule"), i.e. we have:

$$\vec{a} \times \vec{b} = -\vec{b} \times \vec{a}$$

And we cannot simply reduce a vector product. We can reduce the algebraic equation

$$ax = ab$$

by dividing through a. This gives us $x = b$. In the case of the vector product

$$\vec{a} \times \vec{x} = \vec{a} \times \vec{b}$$

we cannot simply assume that $\vec{x} = \vec{b}$ is true. Instead we have:

$$\vec{0} = (\vec{a} \times \vec{b}) - (\vec{a} \times \vec{x}) = \vec{a} \times (\vec{b} - \vec{x})$$

(we can pull out the $\vec{a}$ of the vector product as you can see if you perform the calculation component-by-component.) This equation is zero if either $\vec{a}$ is zero (which would be the trivial solution), or if we have $\vec{b} - \vec{x} = \vec{0}$ (which would be the expected solution $\vec{x} = \vec{b}$), or if $\vec{b} - \vec{x}$ and $\vec{a}$ are (anti-)parallel to each other. Two vectors are (anti-)parallel to each other if they only differ by a scalar multiple t, i.e. the solution has the form

$$\vec{b} - \vec{x} = t\vec{a}$$

Something similar is true for the inner product. We cannot deduce from

$$\vec{a} \cdot \vec{x} = \vec{a} \cdot \vec{b}$$

that we have $\vec{x} = \vec{b}$. Instead we get (similarly to the vector product):

$$0 = \vec{a} \cdot (\vec{b} - \vec{x})$$

The inner product is zero if the vectors $\vec{a}$ and $\vec{b} - \vec{x}$ are orthogonal to each other. But then $\vec{b} - \vec{x}$ could still not be equal to zero and thus $\vec{b} \neq \vec{x}$.

You cannot simply reduce an equation in the case of the inner or vector product. In general, we don't have the inverse operation to the multiplication for vectors. We can divide a vector by a scalar t (simply multiply it with the inverse t^{-1}), but it is not possible to define a division for vectors in a meaningful way.

6.3. Vector Analysis

In physics we generally do not deal with just one scalar or just one vector but with fields which are defined in the whole space, i.e. to each point in space we can assign a scalar or a vector. For example, we can assign the scalar "temperature" to each water molecule in a river. And we can also assign the vector "velocity" to it. We call the field that describes the temperature or the velocity of water molecules a scalar or a vector field, respectively.

Let us take a metal bar and hold one of its ends into a flame. Obviously, this end will be much hotter than the other end. We have a slightly different temperature T at each point x of the metal bar. We have a scalar field $T(x)$ which describes the temperature distribution among the metal bar.

The temperature distribution, however, is not constant. The heat will spread through the metal bar until the whole object has the same temperature. The heat "flows" from the hot end to the cold end. This heat flow has a direction, i.e. it is a vector. Now we would like to know in which direction we have the strongest change of heat. To calculate this, we form the derivative of the scalar function with respect to the three space coordinates and get:

$$\text{grad}\, f = \begin{pmatrix} \frac{\partial f}{\partial x} \\ \frac{\partial f}{\partial y} \\ \frac{\partial f}{\partial z} \end{pmatrix}$$

We have formed the so-called gradient, abbreviated "grad". Often, we also use the Nable-operator ∇ to designate this operation.

$$\nabla f = \begin{pmatrix} \frac{\partial f}{\partial x} \\ \frac{\partial f}{\partial y} \\ \frac{\partial f}{\partial z} \end{pmatrix}$$

The heat flow $\vec{j(x)}$ for a temperature distribution $T(x)$ along the length of the bar is simply:

$$\vec{j(x)} = -\lambda \cdot \nabla T = -\lambda \frac{\partial T}{\partial x}$$

The variable λ is the heat conductance, but we won't go into details here.

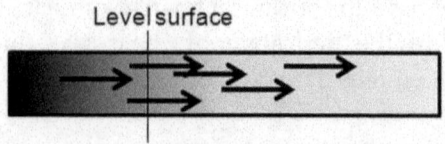

Figure 48: The gradient of a scalar field is always perpendicular to the level surface.

Figure 48 shows an exemplary temperature distribution at a metal bar. The dark grey color stands for a high temperature that gets lower and lower when going to the right. The vectors of the heat flow are parallel to the length of the bar and also show to the right (as the heat flows in this direction). We get the so called level surface if we combine points of the same temperature. In this case, the level surface is perpendicular to the heat flow. But this is also true in general: The level surface is always perpendicular to the gradient.

*

When we have a vector field we might want to know how fast a vector field moves away from a given point; in this case, we are only interested in the speed. The result of the calculation is a scalar which is called the divergence of the vector field.

On the other hand, we might also be interested to see how strongly a vector field moves, i.e. rotates around a given point. In this case, our result is not only a scalar, but also a direction. This vector is called rotation of the vector field.

In the case of the divergence we would like to know how fast a vector field moves at a given point. The change of the field is described by the derivative of the field with respect to the space coordinates. If the vector field has the form

$$\vec{F} = \begin{pmatrix} F_1(x,y,z) \\ F_2(x,y,z) \\ F_3(x,y,z) \end{pmatrix}$$

then the divergence is given by:

$$\text{div}\,\vec{F} = \nabla\vec{F} = \frac{\partial F_1}{\partial x} + \frac{\partial F_2}{\partial y} + \frac{\partial F_3}{\partial z}$$

As in the case of the gradient, we use again the Nabla-operator to describe the operation. To remember the formula for the divergence, we can imagine that the divergence is nothing but the inner product of the Nabla-vector

$$\nabla = \begin{pmatrix} \frac{\partial}{\partial x} \\ \frac{\partial}{\partial y} \\ \frac{\partial}{\partial z} \end{pmatrix}$$

with the vector function $\vec{F}$ which can be written as $\nabla \cdot \vec{F}$. One very simple case is the divergence of the vector field

$$\vec{F} = \begin{pmatrix} x \\ y \\ z \end{pmatrix}$$

because then we simply have:

$$\nabla \cdot \vec{F} = 3$$

If the divergence of a flow at a given point is bigger than zero, this means that more is coming out of this point than going into this point. We have a "source" of the flow. If the divergence is smaller than zero, then more goes in than comes out. We have a "sink" of the flow. If the divergence is zero, than we call the field solenoidal. A water source has a divergence greater than zero, just like a positive point charge, whereas a negative point charge is a sink for the electric field.

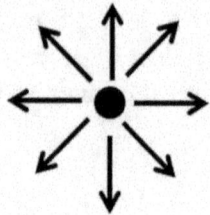

Figure 49: The divergence of a vector field tells us if we have s source (as in this case) or a sink of the vector field.

Figure 49 shows the source of a vector field. The divergence of the point source is greater than zero as the vectors of the vector field (the field lines) all show away from it.

Sometimes we do not only want to know if a vector field has sources and sinks, but also if and how it is moving, mainly if it is rotating around a central point. The movement on a circular line is defined by the axis around which the particle is moving. The rotation of a vector field thus gives us a vector which is parallel to the axis of rotation. The direction is defined similarly to the three-finger rule, and we can imagine the rotation of the vector field $\vec{F}(x,y,z)$ as the vector product of the Nabla-operator with the vector field:

$$\operatorname{rot} \vec{F} = \nabla \times \vec{F} = \begin{pmatrix} \frac{\partial F_3}{\partial y} - \frac{\partial F_2}{\partial z} \\ \frac{\partial F_1}{\partial z} - \frac{\partial F_3}{\partial x} \\ \frac{\partial F_2}{\partial x} - \frac{\partial F_1}{\partial y} \end{pmatrix}$$

Let's take a simple example. The function

$$\vec{F} = \begin{pmatrix} y \\ -x \\ 0 \end{pmatrix}$$

obviously describes a vector in the x-y-plane. At the point of origin, this vector is zero. For increasing positive x-values the y-component of the vector becomes increasingly negative. For increasing y-values the x-components just increases. The vectors thus form a kind of circle as figure 50 shows.

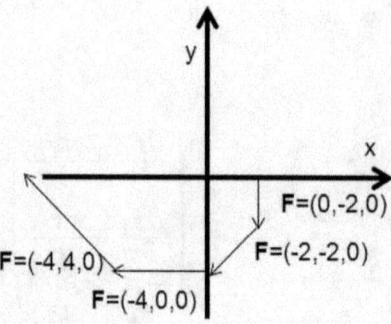

Figure 50: Some vectors of the vector field **F**=(y, -x, 0).

The rotation of the field is:

$$\nabla \times \vec{F} = \begin{pmatrix} 0 \\ 0 \\ -2 \end{pmatrix}$$

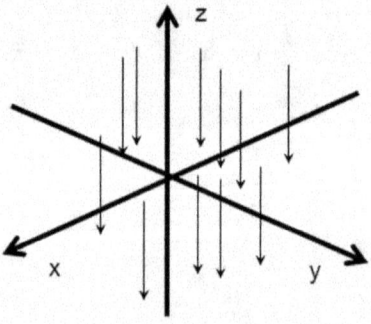

Figure 51: The rotation of the vector field **F**=(y, -x, 0). It has everywhere the constant value of z = -2.

The rotation operation yields the axis of rotation. If the rotation is in the x-y-plane, then the axis of rotation has to be parallel to the z-axis. As the rotation is clockwise, the axis of rotation has to show in the direction of the negative z-axis. And this is also the result of the calculation.

Figure 51 shows some vectors of the rotation $\nabla \times \vec{F}$. They all show into the same direction and they all have the same length.

6.4. Matrices

A function $f(x)$ maps a scalar x onto another scalar y. But just as we can map one scalar onto another scalar, we can also map a vector onto another vector. Let's take the vectors:

$$\vec{x} = \begin{pmatrix} x_1 \\ x_2 \\ x_3 \end{pmatrix}; \; \vec{y} = \begin{pmatrix} y_1 \\ y_2 \end{pmatrix}$$

If we look at a system of equations like

$$y_1 = 2x_1 + 4x_2 - 3x_3$$
$$y_2 = 3x_1 + 2x_3$$

then this system of equations obviously maps the vector $\vec{x}$ onto a vector $\vec{y}$. This is a so-called linear system of equations as the unknowns only appear in their first exponent.

The most important information in this mapping is obviously the set of coefficients of the system of equations. We can summarize them as

$$\begin{pmatrix} 2 & 4 & -3 \\ 3 & 0 & 2 \end{pmatrix}$$

This object is called a matrix. Mapping the vector $\vec{x}$ onto the vector $\vec{y}$ with the help of the matrix A can thus also be written as:

$$\vec{y} = A\vec{x} = \begin{pmatrix} 2 & 4 & -3 \\ 3 & 0 & 2 \end{pmatrix} \vec{x}$$

In general, a matrix made of m-rows and n-columns is an (m, n)-matrix (m by n matrix). Its general form is:

$$A = \begin{pmatrix} a_{11} & \cdots & a_{1n} \\ \vdots & \ddots & \vdots \\ a_{m1} & \cdots & a_{mn} \end{pmatrix}$$

The variables a_{ik} are the elements of the matrix. The element a_{ik} can be found in the ith row and the kth column. The i is therefore also called the row index and the k the column index.

Mapping a vector onto another vector with the use of a matrix has the general form:

$$\vec{y} = A\vec{x}$$

$$\begin{pmatrix} y_1 \\ \vdots \\ y_m \end{pmatrix} = \begin{pmatrix} a_{11} & \cdots & a_{1n} \\ \vdots & \ddots & \vdots \\ a_{m1} & \cdots & a_{mn} \end{pmatrix} \begin{pmatrix} x_1 \\ \vdots \\ x_n \end{pmatrix}$$

As we have seen when deriving the matrix, the ith component of the vector $\vec{y}$ is the result of the inner product of the ith row of the matrix with the vector $\vec{x}$.

$$y_i = (a_{i1}, a_{i2}, \ldots, a_{in}) \cdot \vec{x}$$

$$y_i = a_{i1}x_1 + a_{i2}x_2 + \cdots + a_{in}x_n$$

We call a matrix a square matrix if the number of rows and columns are identical. The elements $a_{11}, a_{22}, \ldots, a_{nn}$ form the so-called main diagonal. If we have a square matrix where all elements apart from the elements of the main diagonal disappear (i.e. are zero) then we have a diagonal matrix. If all elements of a diagonal matrix are ones, then we have a unit matrix. As can be seen directly, a unit matrix maps a vector $\vec{x}$ identically onto the vector $\vec{y}$.

$$\begin{pmatrix} y_1 \\ y_2 \\ y_3 \end{pmatrix} = \begin{pmatrix} 1 & 0 & 0 \\ 0 & 1 & 0 \\ 0 & 0 & 1 \end{pmatrix} \begin{pmatrix} x_1 \\ x_2 \\ x_3 \end{pmatrix}$$

A matrix where all elements are zero is a zero matrix. A $(m, 1)$-matrix is the already known vector. To differentiate it from a $(1, n)$-matrix we also call the well-known vector a column vector and the $(1, n)$-matrix a row vector.

If we exchange the rows and the columns in a matrix, then we get the so-called transpose matrix A^T. For the elements a_{ik} of A and a_{ik}^T of A^T we have:

$$a_{ik} = a_{ki}^T$$

If we have the matrix

$$A = \begin{pmatrix} 1 & 2 \\ 3 & 4 \\ 5 & 6 \end{pmatrix}$$

then the transpose matrix is:

$$A^T = \begin{pmatrix} 1 & 3 & 5 \\ 2 & 4 & 6 \end{pmatrix}$$

We sometimes use the transpose form of writing if, due to space restrictions, we want to place the vector

$$\vec{x} = \begin{pmatrix} 1 \\ 2 \\ 3 \end{pmatrix}$$

in a row and thus write it in the form $\vec{x} = (1, 2, 3)^T$.

We can calculate with matrices as with any other mathematical object, may they be vectors or functions. We can for instance multiply a matrix with a scalar λ. As in the case of the vector, this means that each component of a matrix is multiplied with the scalar, i.e. we have:

$$\lambda A = \begin{pmatrix} \lambda a_{11} & \cdots & \lambda a_{1n} \\ \vdots & \ddots & \vdots \\ \lambda a_{m1} & \cdots & \lambda a_{mn} \end{pmatrix}$$

We can also add two matrices. But to do this, the two matrices must be of the same kind, i.e. they need to have the same number of rows and columns. Then we add them by adding the respective components, i.e.

$$A + B = \begin{pmatrix} a_{11} + b_{11} & \cdots & a_{1n} + b_{1n} \\ \vdots & \ddots & \vdots \\ a_{m1} + b_{m1} & \cdots & a_{mn} + b_{mn} \end{pmatrix}$$

The multiplication of two matrices just means that we perform two mapping operations one after the other. So we first map the vector $\vec{x}$ onto the vector $\vec{y}$ using the matrix B, and then we map $\vec{y}$ onto $\vec{z}$ using the matrix A. We have two mapping operations:

$$\vec{y} = B\vec{x}$$

$$\begin{pmatrix} y_1 \\ y_2 \end{pmatrix} = \begin{pmatrix} b_{11} & b_{12} & b_{13} \\ b_{21} & b_{22} & b_{23} \end{pmatrix} \begin{pmatrix} x_1 \\ x_2 \\ x_3 \end{pmatrix}$$

and

$$\vec{z} = A\vec{y}$$

$$\begin{pmatrix} z_1 \\ z_2 \end{pmatrix} = \begin{pmatrix} a_{11} & a_{12} \\ a_{21} & a_{22} \end{pmatrix} \begin{pmatrix} y_1 \\ y_2 \end{pmatrix}$$

We first calculate the components of the vector $\vec{z}$ and get:

$$z_1 = a_{11} y_1 + a_{12} y_2$$
$$z_2 = a_{21} y_1 + a_{22} y_2$$

Now we replace y_1 and y_2 with the respective values from the first mapping and get:

$$z_1 = a_{11}(b_{11} x_1 + b_{12} x_2 + b_{13} x_3) + a_{12}(b_{21} x_1 + b_{22} x_2 + b_{23} x_3)$$

$$z_2 = a_{21}(b_{11}x_1 + b_{12}x_2 + b_{13}x_3) + a_{22}(b_{21}x_1 + b_{22}x_2 + b_{23}x_3)$$

We perform the multiplication and sort with respect to the x-components. This results in:

$$z_1 = (a_{11}b_{11} + a_{12}b_{21})x_1 + (a_{11}b_{12} + a_{12}b_{22})x_2 + (a_{11}b_{13} + a_{12}b_{23})x_3$$

$$z_2 = (a_{21}b_{11} + a_{22}b_{21})x_1 + (a_{21}b_{12} + a_{22}b_{22})x_2 + (a_{21}b_{13} + a_{22}b_{23})x_3$$

On the matrix-level, this calculation looks like:

$$\vec{z} = A \cdot B \vec{x} = C \vec{x}$$

The components of the product matrix C are the coefficients of the components of the vector $\vec{x}$. We thus get:

$$C = A \cdot B = \begin{pmatrix} a_{11}b_{11} + a_{12}b_{21} & a_{11}b_{12} + a_{12}b_{22} & a_{11}b_{13} + a_{12}b_{23} \\ a_{21}b_{11} + a_{22}b_{21} & a_{21}b_{12} + a_{22}b_{22} & a_{21}b_{13} + a_{22}b_{23} \end{pmatrix}$$

In general we have that the element c_{ik} of the product matrix C is the inner product of the ith row of matrix A with the kth column of the matrix B. We thus have:

$$c_{ik} = a_{i1}b_{1k} + a_{i2}b_{2k} + \cdots + a_{in}b_{nk}$$

or within a matrix:

$$\begin{pmatrix} a_{11} & \cdots & a_{1n} \\ \vdots & \cdots & \vdots \\ a_{i1} & \cdots & a_{in} \\ \vdots & \cdots & \vdots \\ a_{m1} & \cdots & a_{mn} \end{pmatrix} \cdot \begin{pmatrix} b_{11} & \cdots & b_{1k} & \cdots & b_{1s} \\ \vdots & & \vdots & & \vdots \\ b_{n1} & \cdots & b_{nk} & \cdots & b_{ns} \end{pmatrix} = \begin{pmatrix} c_{11} & \cdots & c_{1k} & \cdots & c_{1s} \\ \vdots & & \vdots & & \vdots \\ c_{i1} & \cdots & c_{ik} & \cdots & c_{is} \\ \vdots & & \vdots & & \vdots \\ c_{m1} & \cdots & c_{mk} & \cdots & c_{ms} \end{pmatrix}$$

As you can see, the product of two matrices A and B can only be defined if the number of rows of A is equal to the number of columns of B, i.e. if A is a (m, n)-matrix and B a (n, s)-matrix. Then the product matrix C is a (m, s)-matrix. For a multiplication with the unit matrix E we have:

$$AE = EA = A$$

The associative law is valid for matrices as it is for numbers:

$$(AB)C = A(BC)$$

The same is true for the distributive law:

$$A(B + C) = AB + AC$$
$$(A + B)C = AC + BC$$

However, the commutative law is not valid for matrices in general. We can verify this quickly if we multiply the (m, n)-matrix A with the (n, m)-matrix B. In the case of $A \cdot B$ we get a (m, m)-matrix, and in the case of $B \cdot A$ we get a (n, n)-matrix.

And again we cannot deduce that one of the matrices has been a zero matrix if the result of a multiplication is zero. Let us just take the two matrices

$$A = \begin{pmatrix} 1 & 1 \\ 2 & 2 \end{pmatrix}; B = \begin{pmatrix} -1 & 1 \\ 1 & -1 \end{pmatrix}$$

If we multiply them, we will get a zero matrix, but obviously none of them is a zero matrix.

We can find an inverse matrix A^{-1} to some matrices so that we have:

$$A \cdot A^{-1} = A^{-1} \cdot A = E$$

with the unit matrix E. One necessary condition for the existence of an inverse matrix is that the matrix A is a square matrix (otherwise there is no way the commutative law could work). Furthermore we have that the inversion of an inverse matrix results in the original matrix, i.e.

$$(A^{-1})^{-1} = A$$

The multiplication with a scalar for an inverse matrix is defined as:

$$(\lambda A)^{-1} = \frac{1}{\lambda} A^{-1}$$

The question now is how we can calculated the inverse matrix if the original matrix A is known. We will come back to this question a little bit later. Its answer will be given as the modification of another problem we want to discuss now.

We introduced the matrix as a way to describe a system of equations where some variables x_i are mapped onto the variables y_i. In our example, the mapping was described by the system of equations:

$$y_1 = 2x_1 + 4x_2 - 3x_3$$
$$y_2 = 3x_1 + 2x_3$$

In this case, y_i depends on x_i as you would expect from a mapping. Now let's assume that y_i was not a variable but a constant b_i of the vector $\vec{b}$. Then we would have a system of equations of the form:

$$A\vec{x} = \vec{b}$$

Such a system of equations is called inhomogeneous if $\vec{b} \neq \vec{0}$ and homogeneous if $\vec{b} = \vec{0}$. We can find the solutions $\vec{x}$ of this system of equations by using an algorithm known as Gaussian elimination. As a first step, we write down the expanded coefficient matrix:

$$(A|\vec{b}) = \begin{matrix} a_{11} & \cdots & a_{1n} & b_1 \\ \vdots & & \vdots & \vdots \\ a_{m1} & \cdots & a_{mn} & b_m \end{matrix}$$

Generally, it is written down without the brackets for simplicity reason.

The Gaussian elimination algorithm uses some elementary row operations that do not change the solutions for $\vec{x}$:

- Swap the position of two rows
- Multiply a row by a non-zero scalar
- Add to one row a scalar multiple of another

The goal of the algorithm is to bring the matrix in the so-called row echelon form. This is a form where the matrix has only zeros on the lower left part:

$$\begin{matrix} c_{11} & \cdots & \cdots & c_{1n} & d_1 \\ 0 & c_{22} & \cdots & c_{2n} & \vdots \\ \vdots & \vdots & \ddots & \vdots & \vdots \\ 0 & 0 & 0 & c_{mn} & d_m \end{matrix}$$

In this case, we have the very simple equation

$$c_{nm} x_m = d_m$$

in the last row which can be easily solved for x_m. We introduce this value in the row above and calculate x_{m-1} and so on until we have x_1.

Let us try to solve the following system of equations:

$$\begin{aligned} x_1 - x_2 + 2x_3 &= 3 \\ 2x_1 + 3x_3 &= 6 \\ x_1 + x_2 + 2x_3 &= 4 \end{aligned}$$

We get the following coefficient matrix for this system of equations:

$$\begin{matrix} 1 & -1 & 2 & 3 \\ 2 & 0 & 3 & 6 \\ 1 & 1 & 2 & 4 \end{matrix}$$

If the coefficient matrix began with a zero, then we would swap another row to the top which starts with a number which does not equal zero so that we can achieve the row echelon form. In this case, the first row starts with a one so we don't have to swap rows.

To get the row echelon form, the second and third rows have to begin with a zero. To get this, we have to add or subtract rows. We can see instantly that we get a zero on the first position of the second row if we subtract the double of the first row from the second row. We then have:

$$\begin{array}{cccc} 1 & -1 & 2 & 3 \\ 0 & 2 & -1 & 0 \\ 1 & 1 & 2 & 4 \end{array}$$

We get the zero on the first position of the third row if we subtract the first row from the third:

$$\begin{array}{cccc} 1 & -1 & 2 & 3 \\ 0 & 2 & -1 & 0 \\ 0 & 2 & 0 & 1 \end{array}$$

Now we have to get a zero on the second position of the third row. We can achieve this by subtracting the second row from the third row:

$$\begin{array}{cccc} 1 & -1 & 2 & 3 \\ 0 & 2 & -1 & 0 \\ 0 & 0 & 1 & 1 \end{array}$$

So that we have for x_3:

$$x_3 = 1$$

If we insert this into the second row, we get:

$$2x_2 - 1 \cdot 1 = 0$$

This gives:

$$x_2 = \frac{1}{2}$$

For x_1 we have to calculate:

$$x_1 - 1\cdot\left(\frac{1}{2}\right) + 2\cdot 1 = 3$$

which results in:

$$x_1 = \frac{3}{2}$$

Now we come back to the earlier question of how we can compute the inverse matrix A^{-1} to a given matrix A.
This can also be achieved with the Gaussian elimination algorithm, only in an expanded version which is call Gauss-Jordan algorithm.
We start again with an expanded matrix, only this time we start with the expanded matrix

$$(A|E)$$

with E being the unit matrix. With the operations known from the Gaussian elimination algorithm we bring it into the form

$$(E|A^{-1})$$

Let's do this for our example:

$$\begin{array}{ccccccc} 1 & -1 & 2 & 1 & 0 & 0 \\ 2 & 0 & 3 & 0 & 1 & 0 \\ 1 & 1 & 2 & 0 & 0 & 1 \end{array}$$

Again, we bring the left matrix into the row echelon form. We subtract the double of the first row from the second row to get the new second row and subtract the first row from the third row to get the new third row:

$$\begin{array}{cccccc} 1 & -1 & 2 & 1 & 0 & 0 \\ 0 & 2 & -1 & -2 & 1 & 0 \\ 0 & 2 & 0 & -1 & 0 & 1 \end{array}$$

Then we subtract the second row from the third row:

$$\begin{array}{cccccc} 1 & -1 & 2 & 1 & 0 & 0 \\ 0 & 2 & -1 & -2 & 1 & 0 \\ 0 & 0 & 1 & 1 & -1 & 1 \end{array}$$

Now we have the row echelon form. But we need the diagonal form. To get it, we produce a zero in the last column of the left matrix (of course with the exception of the third row). So we subtract the double of the third row from the first row to get the new first row, and then we add the third row to the second row to get the new second row:

$$\begin{array}{cccccc} 1 & -1 & 0 & -1 & 2 & -2 \\ 0 & 2 & 0 & -1 & 0 & 1 \\ 0 & 0 & 1 & 1 & -1 & 1 \end{array}$$

Now we double the first row and add the second row to get the new first row:

$$\begin{array}{cccccc} 2 & 0 & 0 & -3 & 4 & -3 \\ 0 & 2 & 0 & -1 & 0 & 1 \\ 0 & 0 & 1 & 1 & -1 & 1 \end{array}$$

After all these operations, the left matrix has a diagonal form. But we need a unit matrix, i.e. we have to divide the first and second row by two and get:

$$\begin{matrix} 1 & 0 & 0 & -\frac{3}{2} & 2 & -\frac{3}{2} \\ 0 & 1 & 0 & -\frac{1}{2} & 0 & \frac{1}{2} \\ 0 & 0 & 1 & 1 & -1 & 1 \end{matrix}$$

So the inverse matrix is:

$$A^{-1} = \begin{pmatrix} -\frac{3}{2} & 2 & -\frac{3}{2} \\ -\frac{1}{2} & 0 & \frac{1}{2} \\ 1 & -1 & 1 \end{pmatrix}$$

If we multiply it with the original matrix, we really have $A \cdot A^{-1} = E$.

*

The matrix A maps the vector $\vec{x}$ onto the vector $\vec{y}$. In many cases (especially in physics), we would like to know which vectors do not change its direction after the mapping but are mapped to itself or a multiple λ, i.e. for which we have:

$$A\vec{x} = \lambda\vec{x}$$

The vector $\vec{x}$ which fulfills this equation is called an eigenvector and the scalar λ is called the eigenvalue of the matrix. But before we can compute the eigenvectors and eigenvalues of a matrix, we have to discuss the determinant of a matrix.

The determinant is a scalar which describes the value of a matrix. The (1, 1)-matrix consist only of the element a_{11} which is at the same time the value of the matrix and its determinant.

Describing the determinant of a (2, 2)-matrix (in general, we can only define a determinant for square matrices) is much more difficult to describe. Let's take the matrix A:

$$A = \begin{pmatrix} a_{11} & a_{12} \\ a_{21} & a_{22} \end{pmatrix}$$

This matrix is describe by the two column vectors

$$\vec{a_1} = \begin{pmatrix} a_{11} \\ a_{21} \end{pmatrix} ; \vec{a_2} = \begin{pmatrix} a_{12} \\ a_{22} \end{pmatrix}$$

These vectors describe a parallelogram. Now the value of the matrix is identified with the area of the parallelogram that these two column vectors produce. And we know how we can calculate the area of a parallelogram if we have two vectors: We just calculate the vector product. To do this, we just expand these two vectors by a third dimension and get:

$$\begin{pmatrix} a_{11} \\ a_{21} \\ 0 \end{pmatrix} \times \begin{pmatrix} a_{12} \\ a_{22} \\ 0 \end{pmatrix} = \begin{pmatrix} 0 \\ 0 \\ a_{11}a_{22} - a_{21}a_{12} \end{pmatrix}$$

The value of the vector product and thus the determinant of A is:

$$\det A = a_{11}a_{22} - a_{21}a_{12}$$

The determinant of a (n, n)-matrix can be calculated using Laplace's formula. This formula has the form:

$$\det A = \sum_{j=0}^{n} (-1)^{i+j} \cdot a_{ij} \cdot \det A_{ij}$$

Here A_{ij} is the $(n-1, n-1)$-submatrix of A that we get if we remove the ith row and the jth column.

Let's calculate the determinant

$$\begin{vmatrix} 0 & 1 & 2 \\ 3 & 2 & 1 \\ 1 & 1 & 0 \end{vmatrix}$$

as an example (to differentiate it from the matrix, we use vertical lines as brackets). We develop it with respect to the first row ($i = 1$). We get the first summand by looking at the first column, i.e. $j = 1$. With this we have $(-1)^{i+j} = (-1)^2 = 1$ and $a_{11} = 0$. The subdeterminant A_{ij} is:

$$\begin{vmatrix} 2 & 1 \\ 1 & 0 \end{vmatrix}$$

because we have to remove the first row and the first column. If we do this for all three columns, then we get the sum:

$$\begin{vmatrix} 0 & 1 & 2 \\ 3 & 2 & 1 \\ 1 & 1 & 0 \end{vmatrix} = 0 \cdot \begin{vmatrix} 2 & 1 \\ 1 & 0 \end{vmatrix} - 1 \cdot \begin{vmatrix} 3 & 1 \\ 1 & 0 \end{vmatrix} + 2 \cdot \begin{vmatrix} 3 & 2 \\ 1 & 1 \end{vmatrix}$$

We can calculate the (2, 2)-matrix with the known formula and get:

$$\begin{vmatrix} 0 & 1 & 2 \\ 3 & 2 & 1 \\ 1 & 1 & 0 \end{vmatrix} = 0 + 1 + 2 = 3$$

We can calculate determinants of any size using Laplace's formula. We just have to strip it down until we have the simple to calculate (2, 2) matrices. This means, that the calculation effort for a big matrix can be enormous.

Determinants are mainly used to predict the number of solution for a linear system of equations. Let's take two linear equations of the form:

$$a_{11}x_1 + a_{12}x_2 = b_1$$
$$a_{21}x_1 + a_{22}x_2 = b_2$$

Written in matrix-form we have:

$$A\vec{x} = \vec{b}$$

If we write both equations in the form

$$x_2 = -\frac{a_{11}}{a_{12}}x_1 + \frac{b_1}{a_{12}}$$

$$x_2 = -\frac{a_{21}}{a_{22}}x_1 + \frac{b_2}{a_{22}}$$

then we see that these two equations describe straight lines with the slope of $-\frac{a_{11}}{a_{12}}$ and $-\frac{a_{21}}{a_{22}}$, respectively. Two lines can be positioned in three ways in a plane with respect to each other: They can intersect each other in a point, they can be parallel, or they are identical (i.e. they lie on top of each other), as figure 52 shows.

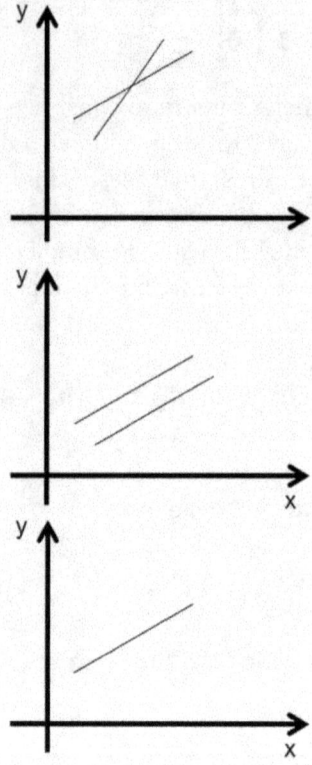

Figure 52: Two straight lines can either intersect in one point, be parallel to each other or even lie on top of each other.

If both lines intersect in one point, then there is exactly one point that both lines have in common, i.e. the system of equations has only one solution.

If both lines are parallel, then there is no solution. If they are identical, then we have an infinite number of solutions. Now we take a look again at the determinant of the matrix A:

$$\det A = a_{11}a_{22} - a_{21}a_{12}$$

If we set the determinant equal to zero, then we get:

$$\frac{a_{11}}{a_{12}} = \frac{a_{21}}{a_{22}}$$

Both sides of this equation are (apart from a minus sign on both sides) equal to the sloped of the lines. If the determinant is zero, then both slopes are identical, i.e. the lines are parallel or even identical. In the first case we have no solution, in the second case an infinite number of solutions.

However, we are looking for just one solution of the system of equations. The lines have to intersect in one point. They can only do this if the slopes are not identical. So we have the result that a linear, inhomogeneous system of equations only has just one solution if the determinant of the coefficient matrix does not equal zero, i.e.

$$\det A \neq 0$$

Is the vector $\vec{b} = \vec{0}$, i.e. if we have a homogeneous system of equations, then both straight lines go through the point of origin and we always have the trivial solution $\vec{x} = \vec{0}$. If we are looking for other solution of the system of equations, then we can only find them if the determinant is $\det A = 0$.

After this short excursus to the determinant, we return to the eigenvalues λ and the eigenvectors $\vec{x}$. We postulated that they have to fulfill the equation

$$A\vec{x} = \lambda\vec{x}$$

We can expand this equation with the unit matrix on the right side and get:

$$A\vec{x} = \lambda E\vec{x}$$

$$A\vec{x} - \lambda E\vec{x} = 0$$

$$(A - \lambda E)\vec{x} = 0$$

As we have just seen, this equation has the trivial solution $\vec{x} = \vec{0}$ if the determinant of the matrix $(A - \lambda E)$ does not equal zero. But we are looking for eigenvectors, i.e. other solutions than just the trivial one. This means that in our case the determinant of the matrix has to be equal to zero:

$$\det(A - \lambda E) = 0$$

This equation is called the characteristic polynomial. It gives us the eigenvalues λ. Using the equation

$$(A - \lambda E)\vec{x} = 0$$

we can then determine the eigenvectors $\vec{x}$.

Let us for example determine the eigenvalues and eigenvectors of the matrix

$$A = \begin{pmatrix} 1 & 4 \\ 1 & -2 \end{pmatrix}$$

In the first step, we will calculate the eigenvalues. To do this, we solve the characteristic polynomial:

$$\det(A - \lambda E) = \det\begin{pmatrix} 1-\lambda & 4 \\ 1 & -2-\lambda \end{pmatrix}$$

$$= (1-\lambda)(-2-\lambda) - 4$$

$$= \lambda^2 + \lambda - 6 = 0$$

We have discussed the formula to solve a quadratic equation earlier. This gives us the eigenvalues of the matrix:

$$\lambda_1 = 2;\ \lambda_2 = -3$$

To get the eigenvectors that belong to these eigenvalues, we have to solve the equation

$$(A - \lambda E)\vec{x} = 0$$

If we insert $\lambda_1 = 2$ into the equation, we get:

$$\begin{pmatrix} 1-2 & 4 \\ 1 & -2-2 \end{pmatrix}\begin{pmatrix} x_1 \\ x_2 \end{pmatrix} = \begin{pmatrix} 0 \\ 0 \end{pmatrix}$$

or

$$\begin{pmatrix} -1 & 4 \\ 1 & -4 \end{pmatrix}\begin{pmatrix} x_1 \\ x_2 \end{pmatrix} = \begin{pmatrix} 0 \\ 0 \end{pmatrix}$$

In fact, we only have to solve one equation:

$$-x_1 + 4x_2 = 0$$

The second equation is identical to this one, only multiplied by -1 (after all, both lines are on top of each other). The equation is thus solved by the vector

$$\vec{x} = \begin{pmatrix} 4 \\ 1 \end{pmatrix}$$

and its multiples (We remember: When looking for the eigenvector, we do not expect to get just one solution but a bunch of vectors that all show into the same direction, but have a different length)

For the second eigenvalue $\lambda_2 = -3$ we get

$$\begin{pmatrix} 1+3 & 4 \\ 1 & -2+3 \end{pmatrix} \begin{pmatrix} x_1 \\ x_2 \end{pmatrix} = \begin{pmatrix} 0 \\ 0 \end{pmatrix}$$

and thus the equation

$$4x_1 + 4x_2 = 0$$

or

$$x_1 + x_2 = 0$$

This equation is solved by

$$\vec{x} = \begin{pmatrix} 1 \\ -1 \end{pmatrix}$$

and its multiples.

We thus have the results that the eigenvalues and eigenvectors of the matrix

$$A = \begin{pmatrix} 1 & 4 \\ 1 & -2 \end{pmatrix}$$

have the form:

$$\lambda_1 = 2; \; \vec{x_1} = a \cdot \begin{pmatrix} 4 \\ 1 \end{pmatrix}$$

and

$$\lambda_2 = -3; \; \vec{x_2} = a \cdot \begin{pmatrix} 1 \\ -1 \end{pmatrix}$$

The number a is any real number that does not equal zero.

7. Fractal Geometry

Fractal geometry is a very young field in mathematics. The term "fractal" was only framed in the year 1975 by the French American mathematician Benoît Mandelbrot who died in 2010. He mainly founded this mathematical field with his work.

Fractal geometry started with the observation that some objects are self-similar. This means that they resemble themselves independent of the magnification that you use to observe them or the details that you look at. Mandelbrot himself was looking at the price of cotton. While doing this he observed that the price history was similar independent of the time period that was chosen – even if an event as the Second World War was in that period.

Later on he was working on noise that disturbed data transmission between computers. Again, this noise hat a random pattern, but the distribution of the noise was self-similar, independent of the scale you were looking at it. It resembled a set that was first described by Georg Cantor in the 19th century. Cantor was one of the founders of the set theory. The set is also known as Cantor dust and is derived iteratively. Figure 53 show the first steps to derive the Cantor dust.

We start with the closed interval $[0,1]$ of real numbers to derive the Cantor dust. The interval is colored black. Then we divide this interval into three equal parts and remove the middle part, i.e. we keep the intervals $[0,\frac{1}{3}]$ and $[\frac{2}{3},1]$. We again remove the middle third of these intervals and get the four intervals $[0, \frac{1}{9}]$, $[\frac{2}{9}, \frac{1}{3}]$, $[\frac{2}{3}, \frac{7}{9}]$, and $[\frac{8}{9}, 1]$. We continue like this, removing the middle part of every interval that remains. The line from zero to one becomes a number of points in this interval. At the limit, the share of the Cantor dust in this interval approaches zero – while having an infinite number of points.

Figure 53: The first iteration steps to derive the Cantor dust.

Before Mandelbrot thought about this, sets like the Cantor dust were considered to be mathematical curiosities. But Mandelbrot wondered what the dimension of this set would be. It is no longer a line – so it can't be one. But it isn't just one point – so zero isn't the right answer, either.

Mandelbrot carried this to the extremes with his famous question: How long is the coast of Britain? He was inspired to ask this question by a work done by the British scientist L. F. Richardson who had discovered differences of up to 20% in different tables about the length of frontiers around countries.

Mandelbrot thought he had a simple explanation for this: It can be explained by the scale that was used.

Let's assume a land surveyor walks along the coast of Britain with a measuring rod of the length of a kilometer. He would not be able to measure small bulges of the coastline; he would take them as straight lines. At the end he notes how often he had to use his measuring rod while going around Britain. This would give him the length of the coastline.

He then repeats the experiment, but this time with a measuring rod of the length of ten meters. Some bulges that he could only ignore with the measuring rod of one kilometer are now measurable. This time he can follow them. He thus has to use the measuring rod more than one hundred times to cover the distance he had covered using his longer measuring rod only once. With even smaller measuring rods allows him to follow even smaller bulges and details of the coast. So the measured length increases with the reduced length of the measuring rod. What, then, is the length of the coast of Britain?

This approach resembles the Koch curve which was presented in 1904 by the Swedish mathematician Helge von Koch. This curve is also derived iteratively as can be seen in figure 54 for the first three steps.

To get the Koch curve, we start with a triangle and add a triangle to the middle of each side which has just one third of the size of the original triangle. Then we add a triangle to each side that has only one third of the size of the triangle that we had used in the last step, and so on.

This resembles measuring a coastline with a large measuring rod, thus ignoring the fine details, and finding more and more details the smaller the length of the measuring rod.

In the end we have a Koch curve with infinite length – but limited to a small area in the plane. So is the dimension of this curve one (after all it is a line) or even two (after all is covers a large area)?

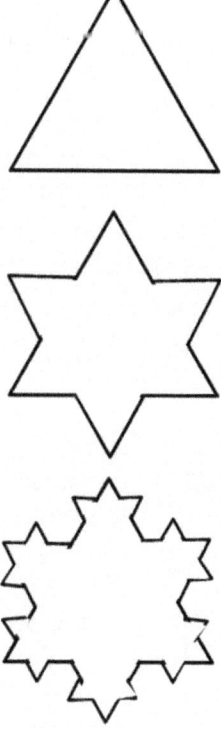

Figure 54: The first three iteration steps to derive the Koch curve.

Again, we are somewhat at a loss with the traditional understanding of a dimension. Mandelbrot therefore concluded that we would need another understanding of dimension, an understanding that would allow using non-integer dimensions.

When trying to find such a definition, he fell back to a definition of dimensions that was made by the German mathematician Felix Hausdorff at the beginning of the 20th century.

Hausdorff proposed to consider the number N of D-dimensional spheres with the radius R that is required to cover the set of points whose dimension we are trying to determine. The Hausdorff dimension D is calculated from the number of spheres that we need when their radius is approaching zero.

The number of spheres depends on the radius, and we need more spheres the smaller the radius is. We have:

$$N(R) \sim \frac{1}{R^D}$$

To find the exponent, we have to find the logarithm (in this case we use the base of ten). At the same time, R is meant to approach zero, i.e. the Hausdorff dimension D is defined as:

$$D = -\lim_{R \to 0} \frac{\log N}{\log R}$$

For a finite and one-dimensional curve the number N of spheres is inversely proportional to the radius of the spheres, i.e. we have:

$$D = -\lim_{R \to 0} \frac{\log \frac{1}{R}}{\log R} = -\lim_{R \to 0} \left(\frac{\log 1 - \log R}{\log R} \right)$$

The logarithm of one is zero, the expression in the brackets gives -1, i.e. we get the Hausdorff dimension: $D = 1$.

For a plane, the number of sphere goes with $\frac{1}{R^2}$, and we get $D = 2$. The Hausdorff definition thus gives us the correct integer dimension for the traditional mathematical objects.

Mandelbrot generalized this definition of dimensions for his fractals. According to him, the dimension is defined as

$$D = \frac{\log(\text{number of self-similar parts})}{\log(\text{reduction ratio})}$$

This definition also gives us the known integer dimensions. For example, we can describe a square as made out of 9 small squares, each one of them one third of the size of the original square (see figure 55).

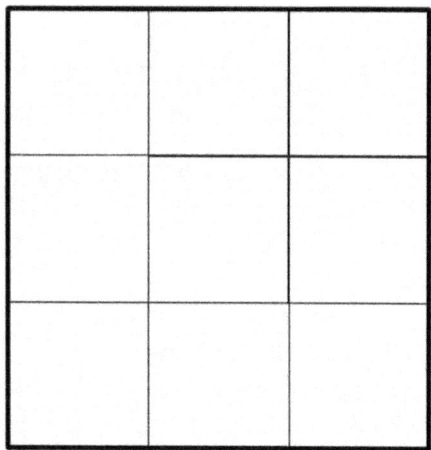

Figure 55: A square made of nine self-similar squares.

This gives us for the dimension:

$$D = \frac{\log 9}{\log 3} = 2$$

The reduction ratio for the Koch curve is three (the triangle put on the sides is one third smaller than the original one), the number of self-similar parts is four as four sides form the repetitive pattern which results in the Koch curve. So the dimension of the Koch curve is:

$$D = \frac{\log 4}{\log 3} = 1.2618...$$

This is a number between one and two as suspected.
The reduction ratio for the Cantor dust is three again, and we have 2 self-similar elements after which the set is repeated, i.e. the fractal dimension of the Cantor dust is:

$$D = \frac{\log 2}{\log 3} = 0.6309...$$

This is smaller than the dimension of a straight line – and bigger than the dimension of a point.
Fractals are not just a mathematical curiosity. We can find them quite often in nature. The blood vessels or bronchia, for instance, have a fractal structure, just like trees, coastlines or plants like the cauliflower.
Fractal geometry also plays an important role in chaos theory, a new field of physics. Chaos theory deals with physical systems like weather or pendulums whose behavior is deterministic and calculable in theory – but can nevertheless not be predicted exactly. For the behavior of the weather or the oscillation of some pendulums depends strongly on the initial conditions of the systems. If they only change slightly, the system will behave completely differently. And as nobody can measure the initial condition of a physical system precisely (no measurement is perfect), this has the consequence that the behavior of chaotic systems like the weather is inherently unpredictable.

Physicists describe the behavior of dynamical system in the so-called "phase space". This is just a coordinate system with each axis representing a state that is used to describe the system. In the case of the pendulum the two coordinates of the phase space are "amplitude" and "speed". In the case of a chaotic system, we get a curve in the phase space that seems to fill up the space without ever doing it completely. The curve resembles a Koch curve, it is a fractal. We need a fractal dimension to describe the curve in the phase space for a chaotic system.

This brings us to the end of the chapter about fractal geometry – without even having mentioned the probably best known fractal, the Mandelbrot set.

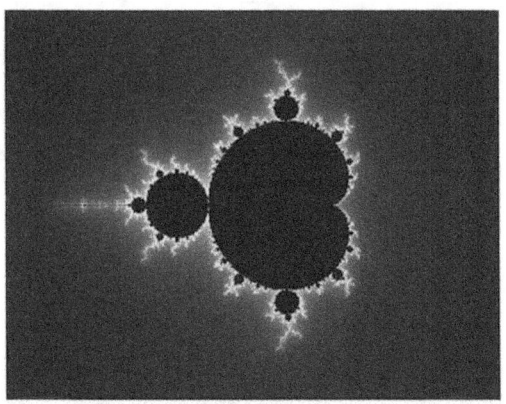

Figure 56: The Mandelbrot set.

Figure 56 shows the well known picture of the Mandelbrot set. You recognize the black core, the colored (here grey) outside space and the heavily jagged border of the set. The set was discovered in 1978 and later on named after Benoît Mandelbrot who wrote a paper about it in 1980.

The picture represents a set of numbers. We are looking from above onto the plane of complex numbers. The point of origin is in the large, black spot. The Mandelbrot set is, strictly speaking, only the black central area of the picture, everything around the border is not part of it.

There is a simple formula to determine if a point of the complex plane is part of the Mandelbrot set (then it is colored in black), or if it isn't. It is a recursive formula:

$$z_{n+1} = z_n^2 + c$$

with the starting point:

$$z_0 = 0$$

We calculate this recursive formula for different complex numbers c and see if it approaches a limit or grows infinitely. If it has a limit, then the point c is part of the Mandelbrot set. All other points are not part of the set and get a different color.

This would give us a rather boring, two color picture. To make it a little bit more interesting, we add a second criterion: If a number is $|z_i| > 2$ then we can be sure that the sequence does not have a limit. The computer stops its calculation and defines this point as not being part of the Mandelbrot set. But depending on how many steps we need to reach this limit for a divergent sequence, we can assign another color to the point. Thus we get the colorful picture for the points that are not part of the Mandelbrot set.

We can see that the Mandelbrot set really is a fractal if we zoom into its borderline as figure 57 shows.

Figure 57: The Mandelbrot set is self-similar.

We can see clearly that the borderline is formed by an endless number of small, bended and self-similar elements.

The calculation for the dimension of this border cannot be done as easily as in the case of the Cantor dust. But estimations show that the dimension of the fractal borderline seems to have, quite ironically, the integer value of two.

With the Mandelbrot set, a set in the complex plane, we close our circle from numbers to fractals.

www.ingramcontent.com/pod-product-compliance
Lightning Source LLC
Chambersburg PA
CBHW051804170526
45167CB00005B/1870